Low Power Methodology Manual
For System-on-Chip Design

Michael Keating • David Flynn •
Robert Aitken • Alan Gibbons • Kaijian Shi

Low Power Methodology Manual

For System-on-Chip Design

ARM Free Library Program

 Springer

Michael Keating
Synopsys, Inc.
Palo Alto, CA
USA

David Flynn
ARM Limited
Cambridge
United Kingdom

Robert Aitken
ARM, Inc.
Almaden, CA
USA

Alan Gibbons
Synopsys, Inc.
Northampton
United Kingdom

Kaijian Shi
Synopsys, Inc.
Dallas, TX
USA

Library of Congress Control Number: 2007928355

ISBN 978-0-387-71818-7   e-ISBN 978-0-387-71819-4

Printed on acid-free paper.

9 8 7 6 5 4 3 2
Corrected at second printing, 2008

springer.com

TRADEMARKS

Synopsys and NanoSim are registered trademarks of Synopsys, Inc.

ARM and AMBA are registered trademarks of ARM Limited. ARM926EJ-S, ARM1176JZF-S, AHB and APB are trademarks of ARM Limited. Artisan and Artisan Components are registered trademarks of ARM Physical IP, Inc.
"ARM" is used to represent ARM Holdings plc; its operating company ARM Limited; and the regional subsidiaries ARM INC.; ARM KK; ARM Korea Ltd.; ARM Taiwan; ARM France SAS; ARM Consulting (Shanghai) Co. Ltd.; ARM Belgium N.V.; AXYS Design Automation Inc.; AXYS GmbH; ARM Embedded Technologies Pvt. Ltd.; and ARM, Inc. and ARM Norway, AS.

All other brands or product names are the property of their respective holders.

DISCLAIMER

All content included in this Low Power Methodology Manual is the result of the combined efforts of ARM Limited and Synopsys, Inc. Because of the possibility of human or mechanical error, neither the authors, ARM Limited, Synopsys, Inc., nor any of their affiliates, including but not limited to Springer Science+Business Media, LLC, guarantees the accuracy, adequacy or completeness of any information contained herein and are not responsible for any errors or omissions, or for the results obtained from the use of such information. THERE ARE NO EXPRESS OR IMPLIED WARRANTIES, INCLUDING, BUT NOT LIMITED TO, WARRANTIES OF MERCHANTABILITY OR FITNESS FOR A PARTICULAR PURPOSE relating to the Low Power Methodology Manual. In no event shall the authors, ARM Limited, Synopsys, Inc., or their affiliates be liable for any indirect, special or consequential damages in connection with the information provided herein.

Table of Contents

Preface

The Low Power Methodology Manual is the outcome of a decade-long collaboration between ARM and Synopsys commercially and the two of us personally. In 1997 ARM and Synopsys worked together to develop a synthesizable ARM7 core. Dave was the ARM lead on the project; Mike's team executed the Synopsys side of the project. This led to a similar project on the ARM9.

Shortly after these projects, the two of us embarked on a series of technology demonstration projects. We both felt that we needed to use our products as our customers do in order to understand how to make these products better. So we developed a test chip that combined ARM and Synopsys IP and took it through to silicon. We did the RTL design and verification personally, and borrowed resources to do the implementation. The experience was incredibly illuminating, and we hope it contributed to improving the IP and tools from both companies.

We quickly realized that low power was one of the key concerns of our customers, and SoC designers in general. So we followed our initial project with several low power technology demonstration projects. The final project was the SALT (Synopsys ARM Low-power Technology demonstrator) project, for which we received working silicon late last year. These projects explored clock gating, multi-voltage, dynamic voltage scaling, and power gating. In all these projects we found that there is no substitute for direct first-hand experience doing low-power IP-based designs. We learned, in the most concrete way possible, exactly what our customers go through on an SoC design.

For years we have been talking about writing a book on low power design. With our experience on the SALT project, our work with customers on low power designs, and our participation in developing the UPF low-power standard, we feel that we are finally in a position to publish our insights and perspectives.

In doing so, we have enlisted the aid of our co-authors. The two of us are primarily front-end engineers, with a background in system architecture and RTL design. Kaijian and Rob bring a great depth of technical expertise in the physical and circuit design aspects of low power. Alan has developed low power flows for the ARM processors and did the implementation of SALT. As a result, he brings a unique perspective on the implementation issues in low power design.

We cannot overstate the contribution of our co-authors. Without their insights and expertise - as well as the material they contributed directly - this book could not have been written.

Like all our joint projects, this book was partly a formal joint project of the two companies and partly (perhaps mostly) driven by the personal commitment of the authors, aided and abetted by many others. We got considerable help from many people for whom this was not part of their job description. These kind souls took time out of their busy schedules, including evenings and weekends, to help us at every step of our journey, from the first joint chip development to the completion of this book. They helped in the architecture, design and tape out of test chips, the building and debugging of boards, and the review and editing of the final manuscript.

It is impossible to list them all, but we list some of the many who contributed to this effort: Anwar Awad, John Biggs, Pin-Hung Chen, Sachin Rai, David Howard, and Sachin Idgunji.

We would also like to thank the staffs of TSMC and UMC for fabricating the technology demonstrators and enabling us to derive the results referenced in the worked examples.

Dave Flynn   Mike Keating
Cambridge, UK   Palo Alto, CA

CHAPTER 1 _____ *Introduction*

1.1 Overview

The design of complex chips has undergone a series of revolutions during the last twenty years. In the 1980s there was the introduction of language-based design and synthesis. In the 1990s, there was the adoption of design reuse and IP as a mainstream design practice. In the last few years, design for low power has started to change again how designers approach complex SoC designs.

Each of these revolutions has been a response to the challenges posed by evolving semiconductor technology. The exponential increase in chip density drove the adoption of language-based design and synthesis, providing a dramatic increase in designer productivity. This approach held Moore's law at bay for a decade or so, but in the era of million gate designs, engineers discovered that there was a limit to how much new RTL could be written for a new chip project. The result was that IP and design reuse became accepted as the only practical way to design large chips with relatively small design teams. Today every SoC design employs substantial IP in order to take advantage of the ever increasing density offered by sub-micron technology.

Deep submicron technology, from 130nm on, poses a new set of design problems. We can now implement tens of millions of gates on a reasonably small die, leading to a power density and total power dissipation that is at the limits of what packaging, cooling, and other infrastructure can support. As technology has shrunk to 90nm and below, the leakage current is increasing dramatically, to the point where, in some 65nm designs, leakage current is nearly as large as dynamic current.

These changes are having a significant effect on how chips are designed. The power density of the highest performance chips has grown to the point where it is no longer possible to increase clock speed as technology shrinks. As a result, designers are designing multi-processor chips instead of chips with a single, ultra-high speed processor.

For battery-powered devices, which comprise one of the fastest growing segments of the electronics market, the leakage of deep submicron processes is a major problem. To combat this problem, designers are using aggressive approaches at every step of the design process, from software to architecture to implementation. These approaches include power gating, where blocks are powered down when not in use, and multi-threshold libraries that can trade-off leakage current for speed.

For all applications, the total power consumption of complex SoCs presents a challenge. To address this challenge, designers are moving from a monolithic approach for power the chip—where a single supply voltage is used for all the non-IO gates of the design—to a multiple supply architecture, where different blocks are run at different voltages, depending on their individual requirements. And in some cases, designers are using voltage scaling techniques to change the supply voltage (and clock frequency) to a critical block depending on its workload and hence required performance.

This book describes a number of the techniques designers can use to reduce the power consumption of complex SoC designs. Our approach is practical, rather than theoretical. We draw heavily upon the experience we have gained in doing a series of technology demonstrator chips over the last several years. We believe the techniques we describe can be used today by chip designers to improve significantly the chips they design.

1.2   Scope of the Problem

Today some of the most powerful microprocessor chips can dissipate 100-150 Watts, for an average power density of 50-75 Watts per square centimeter. Local hot spots on the die can be several times higher than this number.

This power density not only presents packaging and cooling challenges; it also can pose problems for reliability, since the mean time to failure decreases exponentially with temperature. In addition, timing degrades with temperature and leakage increases with temperature.

Historically, the power in the highest performance chips has increased with each new technology node. But because of the issues posed by the power density, the International Technology Roadmap for Semiconductors (ITRS) predicts that the power for these chips will reach a maximum of 198 Watts in 2008; after that, power will remain constant.

Already, the total power consumption of microprocessor chips presents a significant problem for server farms. For these server farms, infrastructure costs (power, cooling) can equal the cost of the computers themselves.

For battery-powered, hand-held devices, the numbers are smaller but the problem just as serious. According to ITRS, battery life for these devices peaked in 2004. Since then, battery life has declined as features have been added faster than power (per feature) has been reduced.

For virtually all applications, reducing the power consumed by SoCs is essential in order to continue to add performance and features and grow these businesses.

Until recently, power has been a second order concern in chip design, following first order issues such as cost, area, and timing. Today, for most SoC designs, the power budget is one of the most important design goals of the project. Exceeding the power budget can be fatal to a project, whether it means moving from a cheap plastic package to an expensive ceramic one, or causing an unacceptably poor reliability due to excessive power density, or failing to meeting the required battery life.

These problems are all expected to get worse as we move to the next technology nodes. The ITRS makes the following predictions:

Table 1-1

Node	90nm	65nm	45nm
Dynamic Power per cm2	1X	1.4X	2X
Static Power per cm2	1X	2.5X	6.5X
Total Power per cm2	1X	2X	4X

Needless to say, many design teams are working very hard to reduce the growth in power below these forecast numbers, since even at 90nm many designs are at the limit of what their customers will accept.

1.3   Power vs. Energy

For battery operated devices, the distinction between power and energy is critical. Figure 1-1 on page 4 illustrates the difference. Power is the instantaneous power in the device. Energy is the area under the curve—the integral of power over time. The power used by a cell phone, for example, varies depending on the what it is doing—

whether it is in standby with the cover closed, or open and the display is powered up, or downloading from the web. The height of the graph in Figure 1-1 shows the power, but it is energy—the area under the curve—that determines battery life.

Power is height (handwritten)

Energy is area (handwritten)

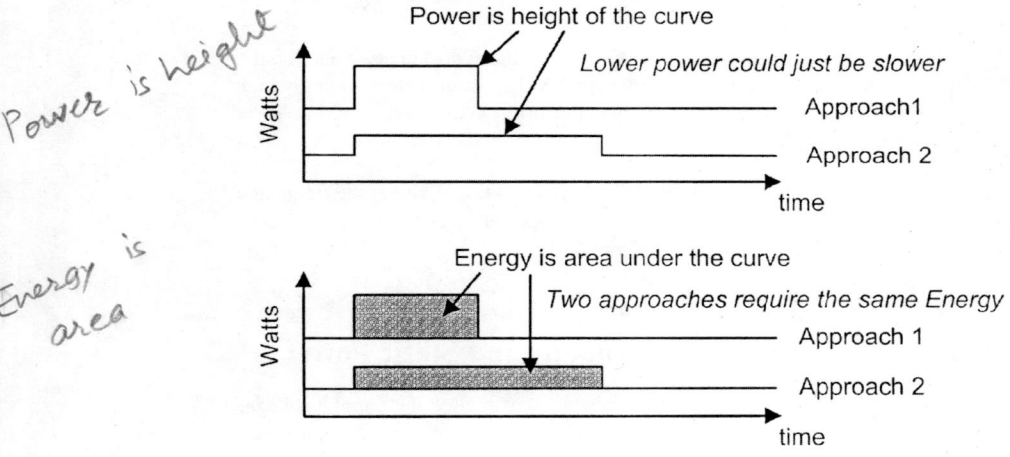

Figure 1-1　Power vs. Energy

1.4　Dynamic Power

The total power for an SoC design consists of dynamic power and static power. Dynamic power is the power consumed when the device is active—that is, when signals are changing values. Static power is the power consumed when the device is powered up but no signals are changing value. In CMOS devices, static power consumption is due to leakage.

The first and primary source of dynamic power consumption is switching power—the power required to charge and discharge the output capacitance on a gate. Figure 1-2 on page 5 illustrates switching power.

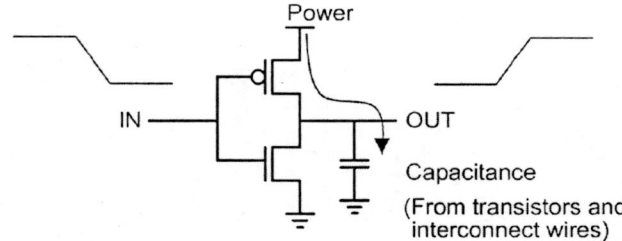

Figure 1-2　Dynamic Power

The energy per transition is given by:

$$Energy\,/\,transition = C_L \bullet V_{dd}^2$$

Where C_L is the load capacitance and V_{dd} is the supply voltage. We can then describe the dynamic power as:

$$P_{dyn} = Energy\,/\,transition \bullet f = C_L \bullet V_{dd}^2 \bullet P_{trans} \bullet f_{clock}$$

Where f is the frequency of transitions, P_{trans} is the probability of an output transition, and f_{clock} is the frequency of the system clock. If we define

$$C_{eff} = P_{trans} \bullet C_L$$

We can also describe the dynamic power with the more familiar expression:

$$P_{dyn} = C_{eff} \bullet V_{dd}^2 \bullet f_{clock}$$

Note that switching power is not a function of transistor size, but rather a function of switching activity and load capacitance. Thus, it is data dependent.

In addition to switching power, internal power also contributes to dynamic power. Figure 1-3 on page 6 shows internal switching currents. Internal power consists of the short circuit currents that occur when both the NMOS and PMOS transistors are on, as well as the current required to charge the internal capacitance of the cell.

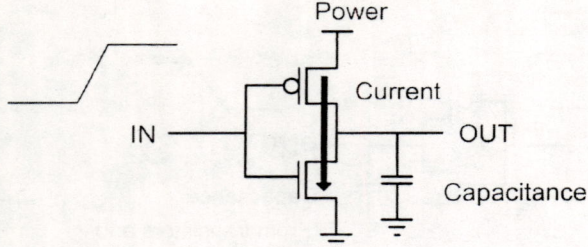

Figure 1-3   Crowbar Current

If we add the expression for internal power to our equation, we can describe the dynamic power as:

$$P_{dyn} = \left(C_{eff} \bullet V_{dd}^2 \bullet f_{clock}\right) + \left(t_{sc} \bullet V_{dd} \bullet I_{peak} \bullet f_{clock}\right)$$

Where t_{sc} is the time duration of the short circuit current, and I_{peak} is the total internal switching current (short circuit current plus the current required to charge the internal capacitance).

As long as the ramp time of the input signal is kept short, the short circuit current occurs for only a short time during each transition, and the overall dynamic power is dominated by the switching power. For this reason, we often simplify the use the switching power formula

$$P_{dyn} = C_{eff} \bullet V_{dd}^2 \bullet f_{clock}$$

But there are occasions when the short circuit current (often called crowbar current) is of interest. In particular, we will discuss ways of preventing excess crowbar current when we talk about how to deal with the floating outputs of a power gated block.

There are a number of techniques at the architectural, logic design, and circuit design that can reduce the power for a particular function implemented in a given technology. These techniques focus on the voltage and frequency components of the equation, as well as reducing the data-dependent switching activity.

There are a variety of architectural and logic design techniques for minimizing switching activity, which effectively lowers switching activity for the gates involved. An interesting example is [1], which describes how engineers have used micro-architecture modifications to reduce power significantly in Intel processors.

Because of the quadratic dependence of power on voltage, decreasing the supply voltage is a highly leveraged way to reduce dynamic power. But because the speed of a gate decreases with decreases in supply voltage, this approach needs to be done carefully. SoC designers can take advantage of this approach in several ways:

- For blocks that do not need to run particularly fast, such as peripherals, we can use a lower voltage supply than other, more speed-critical blocks. This approach is knows as multi-voltage.

- For processors, we can provide a variable supply voltage; during tasks that require peak performance, we can provide a high supply voltage and correspondingly high clock frequency. For tasks that require lower performance, we can provide a lower voltage and slower clock. This approach is known as voltage scaling.

Another approach for lowering dynamic power is clock gating. Driving the frequency to zero drives the power to zero. Some form of clock gating is used on many SoC designs.

1.5   The Conflict Between Dynamic and Static Power

The most effective way to reduce dynamic power is to reduce the supply voltage. Over the last fifteen years, as semiconductor technology has scaled, V_{DD} has been lowered from 5V to 3.3V to 2.5V to 1.2V. The ITRS road map predicts that for 2008 and 2009 high performance devices will use 1.0V and low power devices will use 0.8V.

The trouble with lowering V_{DD} is that it tends to lower I_{DS}, the *on* or drive current of the transistor, resulting in slower speeds. If we ignore velocity saturation and some of the other subtle effects that occur below 90nm, the I_{DS} for a MOSFET can be approximated by:

$$I_{DS} = \mu C_{ox} \frac{W}{L} \cdot \frac{(V_{GS} - V_T)^2}{2}$$

Where μ is the carrier mobility, C_{ox} is the gate capacitance, V_T is the threshold voltage and V_{GS} is the gate-source voltage. From this it is clear that, to maintain good performance, we need to lower V_T as we lower V_{DD} (and hence V_{GS}). However, lowering the threshold voltage (V_T) results in an exponential increase in the sub-threshold leakage current (I_{SUB}), as we show in the following sections.

Thus there is a conflict. To lower dynamic power we lower V_{DD}; to maintain performance we lower V_T; but the result is that we raise leakage current. Until now, this was a reasonable process, since static power from leakage current was so much lower than dynamic power. But with 90nm technology, we are getting to the point where static power can be as big a problem as dynamic power, and we need to examine this conflict more carefully.

1.6   Static Power

There are four main sources of leakage currents in a CMOS gate (Figure 1-4)

- Sub-threshold Leakage (I_{SUB}): the current which flows from the drain to the source current of a transistor operating in the weak inversion region.
- Gate Leakage (I_{GATE}): the current which flows directly from the gate through the oxide to the substrate due to gate oxide tunneling and hot carrier injection.
- Gate Induced Drain Leakage (I_{GIDL}): the current which flows from the drain to the substrate induced by a high field effect in the MOSFET drain caused by a high V_{DG}.
- Reverse Bias Junction Leakage (I_{REV}): caused by minority carrier drift and generation of electron/hole pairs in the depletion regions.

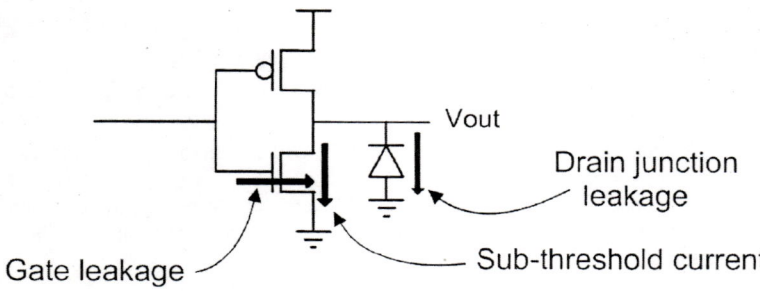

Figure 1-4   Leakage Currents

Sub-threshold leakage occurs when a CMOS gate is not turned completely off. To a good approximation, its value is given by

$$I_{SUB} = \mu C_{ox} V_{th}^2 \frac{W}{L} \cdot e^{\frac{V_{GS}-V_T}{nV_{th}}}.$$

Where W and L are the dimensions of the transistor, and V_{th} is the thermal voltage kT/q (25.9mV at room temperature). The parameter n is a function of the device fabrication process and ranges from 1.0 to 2.5.

This equation tells us that sub-threshold leakage depends exponentially on the difference between V_{GS} and V_T. So as we scale V_{DD} and V_T down (to limit dynamic power) we make leakage power exponentially worse.

Gate leakage occurs as a result of tunneling current through the gate oxide. The gate oxide thickness (T_{OX}) is only a few atoms thick in 90nm gates—this is so thin that tunneling current can become substantial. In previous technology nodes, leakage current has been dominated by sub-threshold leakage. But starting with 90nm, gate leakage can be nearly 1/3 as much as sub-threshold leakage. In 65nm it can equal sub-threshold leakage in some cases. At future nodes, high-k dielectric materials will be required to keep gate leakage in check. This appears to be the only effective way of reducing gate leakage.

Sub-threshold leakage current increases exponentially with temperature. This greatly complicates the problem of designing low power systems. Even if the leakage at room temperature is acceptable, at worst case temperature it can exceed the design goals of the chip.

There are several approaches to minimizing leakage current.

One technique is known as Multi-V_T: using high V_T cells wherever performance goals allow and low V_T cells where necessary to meet timing.

A second technique is to shut down the power supply to a block of logic when it is not active. This approach is known as power gating.

These two approaches are discussed in more detail in later chapters. For now, though, we mention three other techniques:

VTCMOS

Variable Threshold CMOS (VTCMOS) is another very effective way of mitigating standby leakage power. By applying a reverse bias voltage to the substrate, it is possible to reduce the value of the term ($V_{GS}-V_T$), effectively increasing V_T. This approach can reduce the standby leakage by up to three orders of magnitude. However, VTCMOS adds complexity to the library and requires two additional power networks to separately control the voltage applied to the wells. Unfortunately, the effectiveness of reverse body bias has been shown to be decreasing with scaling technology [2].

Stack Effect

The Stack Effect, or self reverse bias, can help to reduce sub-threshold leakage when more than one transistor in the stack is turned off. This is primarily because the small amount of sub-threshold leakage causes the intermediate nodes between the stacked transistors to float away from the power/ground rail. The reduced body-source potential results in a slightly negative gate-source drain voltage. Thus, it reduces the value of the term ($V_{GS}-V_T$), effectively increasing V_T and reducing the sub-threshold leakage. The leakage of a two transistor stack has been shown to be an order of magnitude less than that of a single transistor [3]. This stacking effect makes the leakage of a logic gate highly dependent on its inputs. There is a minimum leakage state for any multi-input circuit; in theory this state applied just prior to halting the clocks to minimize leakage. In practice, applying this state is not feasible in most designs.

Long Channel Devices

From the equation for sub-threshold current, it is clear that using non-minimum length channels will reduce leakage. Unfortunately, long channel devices have lower dynamic current, degrading performance. They are also larger and therefore have greater gate capacitance, which has an adverse effect on dynamic power consumption and further degrades performance. There may not be a reduction in total power dissipation unless the switching activity of the long channel devices is low. Therefore, switching activity and performance goals must be taken in to account when using long channel devices.

1.7   Purpose of This Book

The purpose of the *Low Power Methodology Manual* is to describe the most effective new techniques for managing dynamic and static power in SoC designs. We describe the decisions that engineers need to make in designing low power chips, and provide the information they need to make good decisions. Based on our experience with real chip designs and a set of silicon technology demonstrators, we provide a set of recommendations and describe common pitfalls in doing low power design.

The process of designing a complex chip is itself very complex, involving many stakeholders and participants: systems engineers, RTL designers, IP designers, physical implementation engineers, verification engineers, and library developers. Communication between these disparate players is always a challenge. Each group has its own area of focus, its own priorities, and often its own language. One goal of this book is to give these groups a common language for discussing low power design and a common understanding of the issues involved in implementing a low power strategy.

The first low power decision an SoC design team must make, of course, is what power strategy to pursue—what techniques to use, when and where and on what section of the chip. This fundamental issue drives the structure of the book.

- Chapter 1 (this chapter) gives and over view of the challenges and basic approach to low power design.
- Chapter 2 discusses clock gating methods, Multi-V_T designs, logic-level power reduction techniques, and multi-voltage design.
- Chapter 3 gives a more detailed description of multi-voltage design, focusing on architecture and design issues.
- Chapter 4 gives an overview of power gating
- Chapter 5 addresses design aspects of power gating at the RTL level
- Chapter 6 provides an example of a power gated chip design at the RTL level
- Chapter 7 discusses architectural issues in power gating.
- Chapter 8 discusses issues in IP design for power gating, including an example.
- Chapter 9 discusses architectural and RTL level design issues in dynamic voltage and frequency scaling.
- Chapter 10 discusses some examples of voltage and frequency scaling
- Chapter 11 discusses implementation issues in low power design: synthesis, place and route, timing analysis and power analysis
- Chapter 12 discusses standard cell library and memory requirements for power gating.
- Chapter 13 discusses retention register design and data retention in memories
- Chapter 14 discusses the design of the power switching network
- Appendix A provides some additional information on the circuit design of sleep transistors and power switch networks.
- Appendix B provides detailed descriptions of the UPF commands used in the text.

Throughout the book, we will make reference to several low power technology demonstration projects that the authors have used to explore low power techniques. These projects include:

The SALT project (Synopsys ARM Low power Technology demonstrator) is a 90nm design consisting of an ARM processor and numerous Synopsys peripheral and IO IP. This project focused primarily on power gating techniques. Both the processor and the USB OTG core are power gated.

References

1. Baron, M., "Energy-Efficient Performance at Intel", Microprocessor Report, December 11, 2006.

2. Neau, C. and Roy, K. "Optimal Body Bias Selection for Leakage Improvement and Process Compensation over Different Technology Generations," *Proceedings of the ISLPED, 2003*

3. S. Narendra et al. "Scaling of Stack Effect and its Application for Leakage Reduction", Int. Symp. on Low Power Electronics and Designs, pp.195-200, 2001

CHAPTER 2   *Standard Low Power Methods*

There are a number of power reduction methods that have been used for some time, and which are mature technologies. This chapter describes some of these approaches to low power design.:

- Clock Gating
- Gate Level Power Optimization
- Multi-V_{DD}
- Multi-V_T

2.1   Clock Gating

A significant fraction of the dynamic power in a chip is in the distribution network of the clock. Up to 50% or even more of the dynamic power can be spent in the clock buffers. This result makes intuitive sense since these buffers have the highest toggle rate in the system, there are lots of them, and they often have a high drive strength to minimize clock delay. In addition, the flops receiving the clock dissipate some dynamic power even if the input and output remain the same.

The most common way to reduce this power is to turn clocks off when they are not required. This approach is known as clock gating.

Modern design tools support automatic clock gating: they can identify circuits where clock gating can be inserted without changing the function of the logic. Figure 2-1 shows how this works.

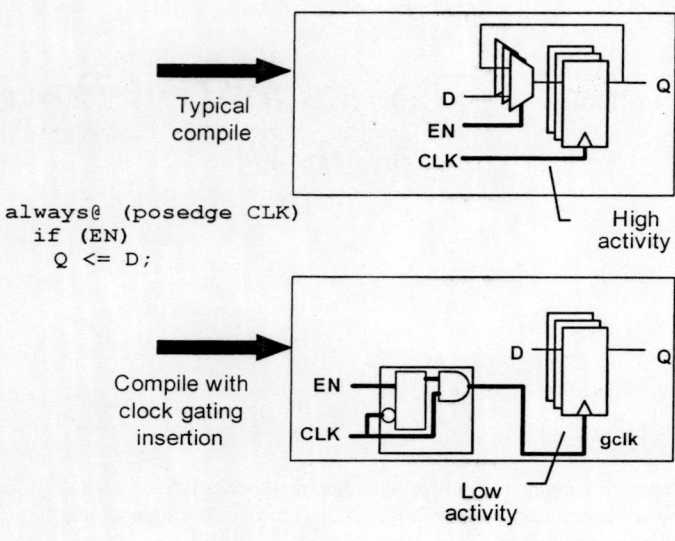

```
always@ (posedge CLK)
  if (EN)
    Q <= D;
```

Figure 2-1   Clock Gating

In the original RTL, the register is updated or not depending on a variable (EN). The same result can be achieve by gating the clock based on the same variable.

If the registers involved are single bits, then a small savings occurs. If they are, say, 32 bit registers, then one clock gating cell can gate the clock to all 32 registers (and any buffers in their clock trees). This can result in considerable power savings.

In the early days of RTL design, engineers would code clock gating circuits explicitly in the RTL. This approach is error prone – it is very easy to create a clock gating circuit that glitches during gating, producing functional errors. Today, most libraries include specific clock gating cells that are recognized by the synthesis tool. The combination of explicit clock gating cells and automatic insertion makes clock gating a simple and reliable way of reducing power. No change to the RTL is required to implement this style of clock gating.

Results

In a recent paper [1], Pokhrel reports on a unique opportunity his team recently had to compare a (nearly) identical chip implemented both with and without clock gating. As a power reduction project, an existing 180nm chip without clock gating was re-

implemented in the same technology with clock gating. Only minor changes in the logic were implemented (some small blocks were removed and replaced by other blocks, for a small net increase in functionality).

Pokhrel reports an area reduction of 20% and a power savings of 34% to 43%, depending on the operating mode. (This savings was realized on the clock gated part of the chip; the processor was a hard macro and not clock gated. Power measurements were made on the whole chip when the processor was in IDLE mode; that is, the processor was turned off.) The power measurements are from actual silicon.

The area savings is due to the fact that a single clock gating cell takes the place of multiple muxes.

Pokhrel makes a couple of interesting observations:

- After some analysis and experiments, the team decided to use clock gating only on registers with a bit-width of at least three. They found that clock gating on one-bit registers was not power or area efficient.
- Much of the power savings was due to the fact that the clock gating cells were placed early in the clock path. Approximately 60% of the clock buffers came after the clock gating cell, and so had their activity reduce to zero during gating.

2.2   Gate Level Power Optimization

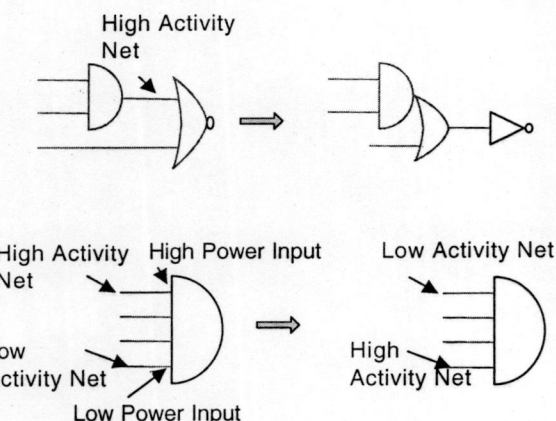

Figure 2-2   Examples of Gate Level Optimizations

In addition to clock gating, there are a number of logic optimizations that the tools can perform to minimize dynamic power. Figure 2-2 shows two of these optimizations.

At the top of the figure, an AND gate output has a particularly high activity. Because it is followed by a NOR gate, it is possible to re-map the two gates to an AND-OR gate plus an inverter, so the high activity net becomes internal to the cell. Now the high activity node (the output of the AND gate) is driving a much smaller capacitance, reducing dynamic power.

At the bottom of the figure, an AND gate has been initially mapped so that a high activity net is connected to a high power input pin, and a low activity net has been mapped to a low power pin. For multiple input gates there can be a significant difference in the input capacitance - and hence the power - for different pins. By remapping the inputs so the high activity net is connected to the low power input, the optimization tool can reduce dynamic power.

Other examples of gate level power optimization include cell sizing and buffer insertion. In cell sizing, the tool can selectively increase and decrease cell drive strength throughout the critical path to achieve timing and then reduce dynamic power to a minimum.

In buffer insertion, the tool can insert buffers rather than increasing the drive strength of the gate itself. If done in the right situations, this can result in lower power.

Like clock gating, gate level power optimization is performed by the implementation tools, and is transparent to the RTL designer.

2.3   Multi V_{DD}

Since dynamic power is proportional to V_{DD}^2, lowering V_{DD} on selected blocks helps reduce power significantly. Unfortunately, lowering the voltage also increases the delay of the gates in the design.

Consider the example in Figure 2-3. Here the cache RAMS are run at the highest voltage because they are on the critical timing path. The CPU is run at a high voltage because its performance determines system performance. But it can be run at a slightly lower voltage than the cache and still have the overall CPU subsystem performance determined by the cache speed. The rest of the chip can run at a lower voltage

still without impacting overall system performance. Often the rest of the chip is running at a much lower frequency than the CPU as well.

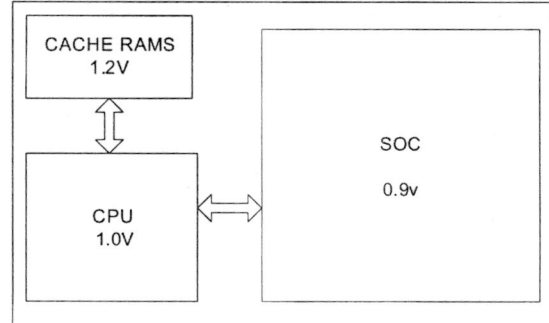

Figure 2-3   Multi-Voltage Architecture

Thus, each major component of the system is running at the lowest voltage consistent with meeting system timing. This approach can provide significant savings in power.

Mixing blocks at different V_{DD} supplies adds some complexity to the design – not only do we need to add IO pins to supply the different power rails, but we also need a more complex power grid and level shifters on signals running between blocks. These issues are described in more detail later in the book.

2.4   Multi-Threshold Logic

As geometries have shrunk to 130nm, 90nm, and below, using libraries with multiple V_T has become a common way of reducing leakage current.

Figure 2-4 shows the relationship between delay and leakage for a 90nm process. Figure 2-5 shows some representative curves for leakage vs. delay for a multi-V_T library. As explained earlier, sub-threshold leakage depends exponentially on V_T. Delay has a much weaker dependence on V_T.

Many libraries today offer two or three versions of their cells: Low V_T, Standard V_T, and High V_T. The implementation tools can take advantage of these libraries to optimize timing and power simultaneously.

leakage library first and then relaxing back any cells not on the critical path by swapping them for their lower performing, lower leakage equivalents.

If minimizing leakage is more important than achieving a minimum performance then this process can be done the other way around: we can target the low leakage library first and then swap in higher performing, high leakage equivalents in speed critical areas.

2.5 Summary of the Impact of Standard Low Power Techniques

Table 2-1 provides a brief summary of the cost/benefit of the techniques described in this chapter.

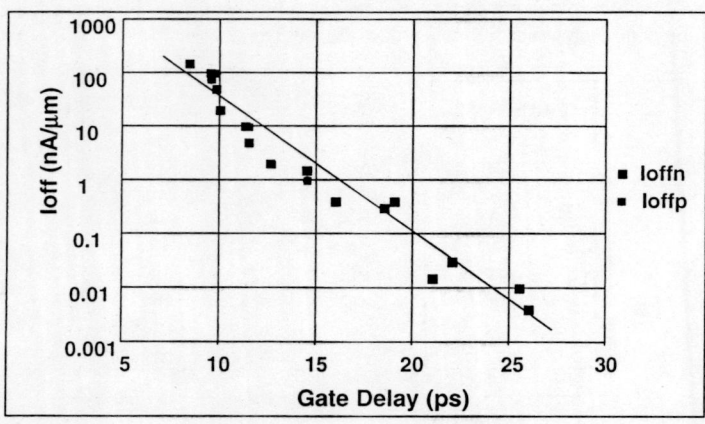

Figure 2-4   Delay vs. Leakage for 90nm

Table 2-1

Tech-nique	Power Benefit	Timing Penalty	Area Penalty	Impact: Architec-ture	Impact: Design	Impact: Verifica-tion	Impact: Place & Route
Multi Vt	Medium	Little	Little	Low	Low	None	Low
Clock Gating	Medium	Little	Little	Low	Low	None	Low
Multi Voltage	Large	Some	Little	High	Medium	Low	Medium

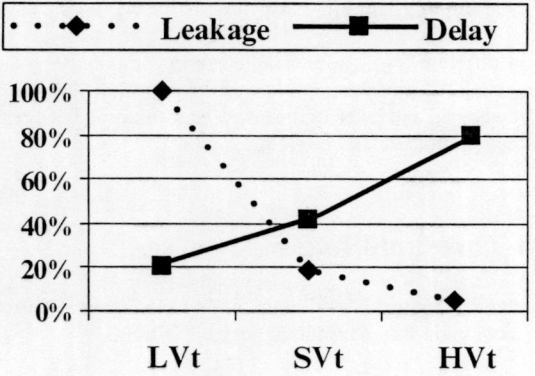

Figure 2-5   Leakage vs. Delay for a 90nm Library

It is now quite common to use a "Dual V_T" flow during synthesis. The goal of this approach is to minimize the total number of fast, leaky low V_T transistors by deploying them only when required to meet timing. This usually involves an initial synthesis targeting a primary library followed by an optimization step targeting one (or more) additional libraries with differing thresholds.

Usually there is a minimum performance which must be met before optimizing power. In practice this usually means synthesizing with the high performance, high

References

1. Pokhrel, K. "Physical and Silicon Measures of Low Power Clock Gating Success: An Apple to Apple Case Study", SNUG, 2007 http://www.snug-universal.org/cgi-bin/search/search.cgi?San+Jose,+2007.

CHAPTER 3

Multi-Voltage Design

The techniques discussed in the previous chapter are mature; engineers have been using them for some time, and design tools have supported them for years. With this chapter, we begin discussing more recent and aggressive approaches to reducing power: power gating and adaptive voltage scaling.

Both of these techniques rely on moving away from the traditional approach of using a single, fixed supply rail for all of the (internal) gates in a design. (IO cells have had a separate power supply in most chips for many years).

The most basic form of this new approach is to partition the internal logic of the chip into multiple voltage regions or power domains, each with its own supply. This approach is called Multi-Voltage design. It is based on the realization that in a modern SoC design, different blocks have different performance objectives and constraints. A processor, for instance, may need to run as fast as the semiconductor technology will allow. In this case, a relatively high supply voltage is required. A USB block, on the other hand, may run at a fixed, relatively low frequency dictated more by the protocol than the underlying technology. In this case, a lower supply rail may be sufficient for the block to meet its timing constraints – and a lower supply rail means that its dynamic and static power will be lower.

Once we have crossed the conceptual barrier of having separate supplies, there are more complex power strategies we can contemplate: we can provide different voltages to our processor, for example, depending on its workload. Or we can provide different voltages to a RAM - a low voltage to maintain memory contents when the memory is not being accessed, and a higher voltage that supports reads and writes. We can even consider dropping the supply voltage to zero – that is, power gating.

For the sake of discussion we provide the following categorization of multi-voltage strategies:

- Static Voltage Scaling (SVS): different blocks or subsystems are given different, fixed supply voltages.
- Multi-level Voltage Scaling (MVS): an extension of the static voltage scaling case where a block or subsystem is switched between two or more voltage levels. Only a few, fixed, discrete levels are supported for different operating modes.
- Dynamic Voltage and Frequency Scaling (DVFS): an extension of MVS where a larger number of voltage levels are dynamically switched to follow changing workloads.
- Adaptive Voltage Scaling (AVS): an extension of DVFS where a control loop is used to adjust the voltage.

3.1   Challenges in Multi-Voltage Designs

Even the simplest multi-voltage design presents the designer with some basic challenges:

- Level shifters. Signals that go between blocks that use different power rails often require level shifters – buffers that translate the signal from one voltage swing to another.
- Characterization and STA. With a single supply for the entire chip, timing analysis can be done at a single performance point. The libraries are characterized for this point, and the tools perform the analysis in a straight-forward manner. With multiple blocks running at different voltages, and with libraries that may not be characterized at the exact voltage we are using, timing analysis becomes much more complex.
- Floor planning, power planning, grids. Multiple power domains require more careful and detailed floorplanning. The power grids become more complex.
- Board level issues. Multi-voltage designs require additional resources on the board – additional regulators to provide the additional supplies.
- Power up and power down sequencing. There may be a required sequence for powering up the design in order to avoid deadlock.

3.2   Voltage Scaling Interfaces – Level Shifters

When driving signals between power domains with radically different power rails, the need for level shifters is clear. Driving a signal from a 1V domain to a 5V domain is a problem – the 1V swing may not even reach threshold in the 5V domain. But the internal voltages in today's chips are tightly clustered around 1V. Why would we need level shifters on signals going from a 0.9V domain to a 1.2V domain?

One fundamental reason is that a 0.9V signal driving a 1.2V gate will turn on both the NMOS and PMOS networks, causing crowbar currents. This issue is discussed later in this chapter.

In addition, standard cell libraries are characterized for – and operate best with – a clean, fast input that goes rail to rail. Failure to meet this requirement may result in signals exhibiting significant rise- or fall-time degradation between the driver cell in one domain and the receiver in another voltage domain. This in turn can lead to timing closure problems and even excessive crowbar switching currents.

The best solution is to make sure each domain gets the voltage swings (and rise- and fall-times) that it expects. We do this by providing level shifters between any domains that use different voltages. This approach limits any voltage swing and timing characterization issues to the boundary of voltage domains, and leaves the internal timing of the domain unaffected. This kind of clean interfacing makes timing closure – and reuse – much easier.

3.2.1   Unidirectional Level Shifters

The design of a level shifter to provide an effective voltage swing between one different voltage rails is an analog design problem. And for analog design reasons, these cells are typically only designed to shift one direction - either from a higher voltage to a lower one, or from a lower voltage to a higher one. Later in this chapter we provide some example designs that show the difference between the two types of cells.

For static voltage scaling, this limitation on level shifters is not a problem. But for the other forms of multi-voltage, where supply voltages can change during operation, it does pose a challenge. The designer must architect and partition the design such that voltage domains have a defined relation to neighboring domains – such as "always higher", "always lower", or "always the same." With this restriction, it then becomes straightforward to implement the interface with the appropriate level shifting components.

Designing interfaces that can operate in both directions may appear attractive from a system perspective but requires non-standard implementation components and tooling.

3.2.2   Level Shifters – High to Low Voltage Translation

On the face of it, simply overdriving a CMOS input from an output buffer on a higher voltage rail does not appear to be a problem – there are no latch-up or breakdown issues, simply a "better", faster edge compared to normal CMOS logic high or low level switching levels.

why have
H→L level shifters?

However for safe timing closure one does need some specially identified "down-shift" cells characterized specifically for this purpose.

If specialized high-to-low level shifter cells were not provided in the library then the entire library would have to be re-characterized to allow accurate static timing analysis. Each gate would have to be characterized for an arbitrary input voltage swing.

As shown in Figure 3-1, high to low level shifters can be quite simple, essentially two inverters in series. Level shifter design is described in more detail in a later chapter, but for now we just observe that require only a single power rail, which is the one from the lower or destination power domain.

As implied by the drawing, a high-to-low level shifter only introduces a buffer delay, so its impact on timing is small.

Figure 3-1   High to Low Level Shifters

3.2.3   Level Shifters – Low-to-High Voltage

Driving logic signals from a low supply rail to a cell on a higher voltage rail is a more critical problem. An under-driven signal degrades the rise and fall times at the receiving inputs. This in turn can lead to higher switching currents and reduced noise margins. A slow transition time means that the signal spends more time near V_T, causing the short circuit (crowbar) current to last longer than necessary.

For clock tree buffering this becomes particularly important. Clock tree buffering is always a challenge, and any degradation in rise and fall times across voltage region boundaries can increase clock skew.

Specially designed level shifter cells solve this problem. They provide fast, full-rail signals to the higher voltage domain. They can be correctly modeled with the design tools to achieve accurate timing.

There are a number of design techniques – but a simple straight-forward design is shown in Figure 3-2. This design takes a buffered and an inverted form of the lower voltage signal and uses this to drive a cross-coupled transistor structure running at the higher voltage.

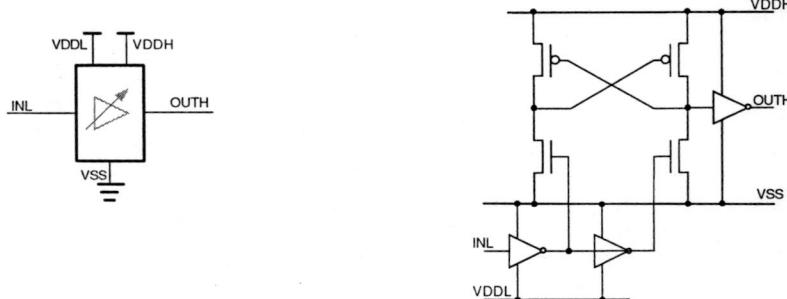

Figure 3-2   Low to High Level Shifters

Such "up-shifting" level converters require two supply rails – and typically share a common ground. The well structures cannot be joined together but must be associated with the supplies independently.

These specialized low to high level shifter cells are characterized over an extended voltage range to match the operating points of both the high side and low side voltage domains. This enables accurate static timing analysis between different voltages and operating conditions.

Low-to-high level shifters introduce a significant delay compared to the simple buffer delays of high-to-low level shifters. In the case of wide interfaces between timing critical blocks – for example, between a CPU and cache memory on different voltage supplies - the designer must take account of the interface delays and any physical routing constraints imposed across the voltage boundary.

3.2.4   Level Shifter Placement

Multi-voltage designs present significant challenges in placement. Figure 3-3 shows an example of two voltage domains embedded in a third voltage domain.

Figure 3-4 shows one possible solution. Here the buffer uses the power rail of the 1.2V domain. But this means that the 1.2V rail must be routed – probably as a signal wire – in the 1.1V domain. This kind of complex power routing is one of the key challenges in automating the implementation of multi-voltage designs.

Figure 3-5 shows the case of a signal from the 0.9V domain going to the 1.2V domain. In this case, power routing will be a challenge no matter where the level shifter is placed. Because it requires both rails, at least one of the rails will have to be routed from another domain. Since the output driver requires more current than the input stage, we place the level shifter in the 1.2V domain.

As with down-shifters, if the distance between the 1.2V domain and the 0.9V domain is small enough, and the library has a strong enough buffer, then the driving buffer can be placed in the 0.9V domain. No additional buffering is required. Otherwise, additional buffers need to be placed in the 1.1V domain, causing the power routing problems mentioned above.

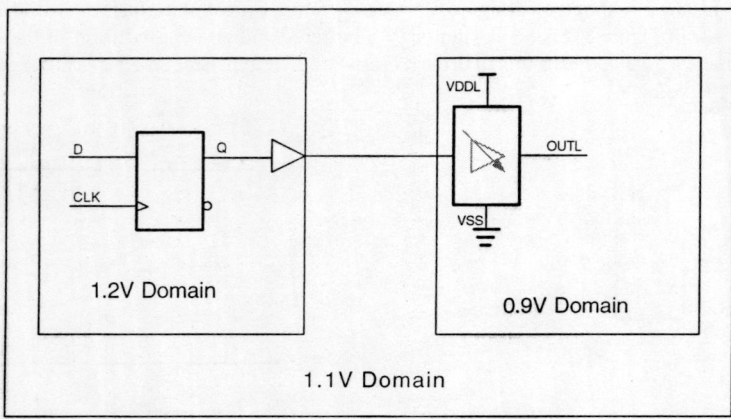

Figure 3-3   Level Shifter in the Destination Domain

Because it uses the voltage rail from the lower voltage domain, the high-to-low level shifter is usually placed in the lower voltage domain. If the distance between the 1.2V domain and the 0.9V domain is small enough, and the library has a strong enough buffer, then the driving buffer can be placed in the 1.2V domain. No additional buffering is required.

Adding additional buffers in the 1.1V domain clearly presents problems – what supply do the buffers use?

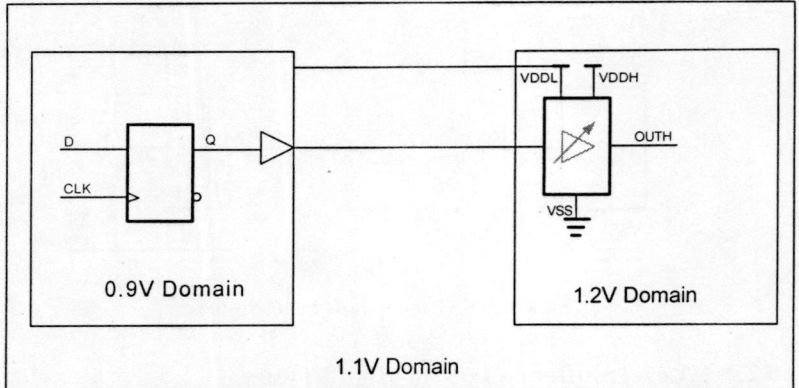

Figure 3-5   Placement of Low to High Level Shifter

3.2.5   Automation and Level Shifters

Level shifters do not affect the functionality of the design; from a logical perspective they are just buffers. For this reason, modern implementation tools can automatically insert level shifters where they are needed. No change to the RTL is required.

Many tools now allow the designer to specify a level shifter placement strategy – to place the low-to-high level shifters in the lower domain, the higher domain, or between them. Note that the output driver has the higher supply current requirements; the low voltage supply only has to power the weaker devices to control the cell. For

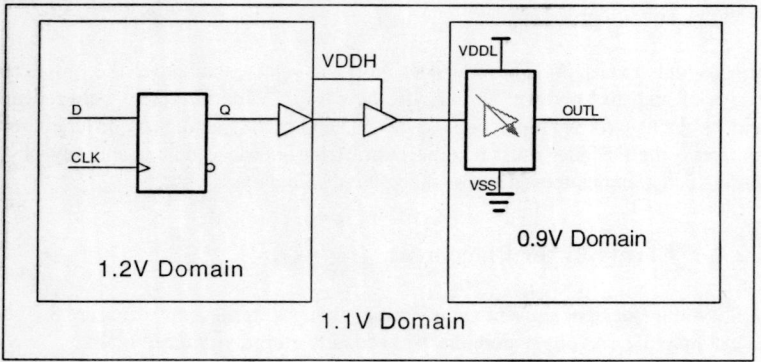

Figure 3-4   Buffering and Level Shifters

this reason we recommend placing the level shifters in the destination domain, as shown in Figure 3-3 and Figure 3-5.

As part of defining a level shifter strategy, the designer specifies rules for when level shifters are inserted. The designer can specify explicitly which blocks require level shifters, or the designer can specify a minimum voltage difference that requires level shifter insertion.

High-to-low level shifters should be inserted based on timing considerations. Using standard gates rather than level shifters at the interface of two different voltage regions causes an error in delay calculation, as mentioned above. If the voltage difference between the two domains is large enough then this timing error becomes unacceptable. In this case, level shifters are required. The exact voltage difference then depends on the library and the design objectives.

Low-to-high level shifters should be inserted based on power as well as timing considerations. If the voltage difference between two domains is large enough, the input stage of a standard gate in the higher domain will not turn all the way off, leading to excessive crowbar current.

Specifically, if the voltage difference is larger than the threshold voltage of the receiving PMOS transistor, the transistor will not completely turn off. In order to provide a reasonable noise margin, we should pad this number by 10% of the VDDH (the higher supply voltage). Thus, if

$$VDDH - VDDL > V_{TPMOS} - (0.1 * VDDH)$$

Then a level shifter should be used in order to shut off the receiving PMOS input transistor stage.

(Here V_{TPMOS} is the threshold voltage of the PMOS transistor, and VDDH and VDDL are the VDD supplies for the higher and lower domains respectively).

3.2.6   Level Shifter Recommendations and Pitfalls

Recommendations:

- Place the level shifters in the receiving domain – in the lower domain for High-to-Low shifters, in the higher domain for Low-to-High shifters.
- Low-to-High level shifters have significant delays that need to be understood and thoughtfully factored into RTL design partitioning for timing critical blocks.
- Ensure there is a defined relationship between different voltage domains such that the operating conditions make it clear whether an up- or down-shifter is required.

Pitfalls:

- Interfaces between domains that may both be higher or lower voltage with respect to each other will require specialized level shifter components and make the setup and hold timing verification across such interfaces very complex.

3.3   Timing Issues in Multi-Voltage Designs

3.3.1   Clocks

Routing clocks across different power domains means that they have to go through level shifters. This clearly complicates automation – the clock tree synthesis tools need to understand level shifters and automatically insert them in the appropriate places.

With multiple level voltage scaling, clock distribution gets even more complex. Consider Figure 3-6. The clock buffers in the multi-level domain will sometimes be powered at 0.9V and sometime at 1.1V. Under which conditions do we attempt to minimize clock skew relative to the clock in the 1.2V domain?

The solution is that optimization and timing analysis must be done simultaneously for both situations, to assure that timing will be met for both conditions.

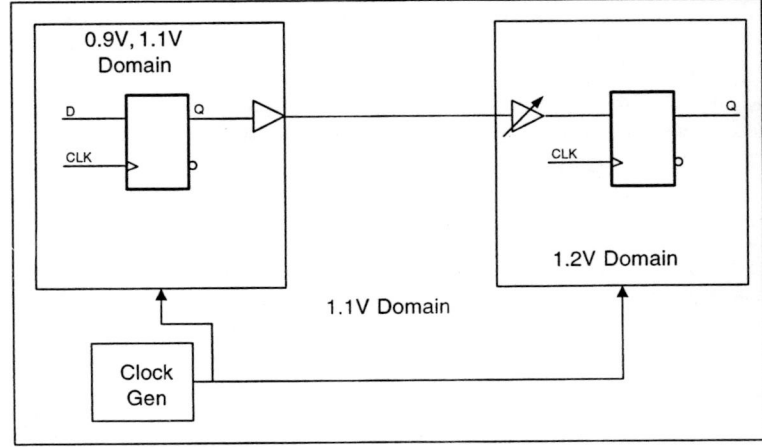

Figure 3-6   Clock Distribution and Multi-Voltage

3.3.2 Static Timing Analysis

In the case of static voltage scaling, timing analysis is not too much more complex than for a single voltage design. It merely requires a library that is characterized for the different voltages begin used. Then the implementation and analysis tools can execute using the appropriate timing information. In the early days of multi-voltage design this was a problem; most tools and libraries assumed a fixed, constant supply voltage for the entire design. But today, this problem has been solved, and static voltage scaling poses little problem for static timing analysis.

Multi-level voltage scaling presents a greater challenge. Again consider Figure 3-6; at which voltage do you do synthesis, place and route, and STA for the multi level block?

The solution is that the timing constraints must be specified for each operating point or supply voltage level. In our example, we must provide two sets of timing constraints for the multi level block, one for 0.9V and one for 1.1V. They may be different because there may be two different operation modes, one for each voltage level, which may have different performance objectives or different clock speeds.

The tools must then perform implementation simultaneously at both 0.9V and 1.1V using these two sets of timing constraints. The implementation is complete only when it meets both sets of requirements with the same implementation.

3.4 Power Planning for Multi-Voltage Design

Just getting power to the different power domains can be a challenge for designs which use multiple supplies. Every voltage scaled region requires an independent local power supply grid, and a low impedance power connection to supply pads.

For flip-chip designs, this problem is mitigated somewhat because power can be delivered locally by a pad located in the power domain. For traditional chips, where the power has to come from the chip periphery, the system designer may need to restrict the number of voltage regions to those that provide significant dynamic power and energy savings.

Power planning for multi-voltage designs is discusses in more detail in Chapter 11.

3.5 System Design Issues with Multi-Voltage Designs

For static voltage scaling, the major implementation issues have to do with level shifters, as discussed above.

The main system level issue is that of power sequencing. In most cases, it will not be practical to bring up all the different power supplies at precisely the same time. Thus, it may be useful to plan an explicit power sequence, so that the different power domains come up in a well-defined order that assures correct function. And in fact, some IP may require a specific power up sequence.

In particular, we need to make sure all power domains are completely powered up before issuing reset. Also, the CPU(s) may need to wait until the rest of the chip is powered up before booting.

Power-on is a particularly complex case because crystal oscillators and Phase-Locked-Loops require technology-dependent stabilization and lock times. These stabilization times only begin once the IO and SOC power supplies are settled.

A power-on-reset Schmitt circuit is one way to guarantee the initial power-up to the SOC is complete. Then some form of timer can be used to determine when the PLL and clocks are stable. Finally, an explicit handshake protocol can be enabled to manage more complex DVFS power management.

Multi-level voltage scaling designs have the additional constraint that ramp times must be carefully controlled to avoid voltage overshoot or undershoot. Since voltages are often changed while the system is running, the system may malfunction or lock up if the voltage is raised above the target voltage or falls significantly below it. This ramp control is best achieved by using a signaling interface to sequence both initial turn-on and subsequent ramping of the voltage regulator.

Finally, the power controller is often controlled by a CPU, which means that power control software must be integrated with the other system software running on the CPU.

CHAPTER 4

Power Gating Overview

Leakage power dissipation grows with every generation of CMOS process technology. This leakage power is not only a serious challenge to battery powered or portable products but increasingly an issue that has to be addressed in tethered equipment such as servers, routers, and set-top boxes.

To reduce the overall leakage power of the chip, it is highly desirable to add mechanisms to turn off blocks that are not being used. This technique is known as power gating.

Section two describes power gating from an RTL design perspective. This chapter provides an overview of power gating. The following chapters continue with descriptions of how to implement power gating at the RTL level, the power gating strategies used on the SALT chip, and the architectural implications of power gating. Our focus is how RTL designers can design power gating implementations in as technology-independent and portable a manner as possible.

4.1 Dynamic and Leakage Power Profiles

The basic strategy of power gating is to provide two power modes: a low power mode and an active mode. The goal is to switch between these modes at the appropriate time and in the appropriate manner to maximize power savings while minimizing the impact to performance.

The power reduction techniques described in chapter 2 do not affect the functionality of the design and do not require changes to the RTL. They can be handled fairly transparently from a design and implementation and perspective; power gating is more invasive than clock-gating in that it affects inter-block interface communication and adds significant time delays to safely enter and exit power gated modes.

Shutting down power to a block of logic may be scheduled explicitly by control software as part of device drivers or operating system idle tasks. Alternatively it may be initiated in hardware by timers or system level power management controllers. In any event, we are faced with architectural trade-offs between

- the amount of leakage power savings that is possible
- the entry and exit time penalties incurred
- the energy dissipated entering and leaving such leakage saving modes
- the activity profile (proportion and frequency of times asleep or active)

First, we introduce some terminology for the entry and exit from power modes:

SLEEP events initiate entry to the low power mode

WAKE events initiate return to active mode

Figure 4-1 shows an example activity profile for a sub-system using clock gating to reduce power.

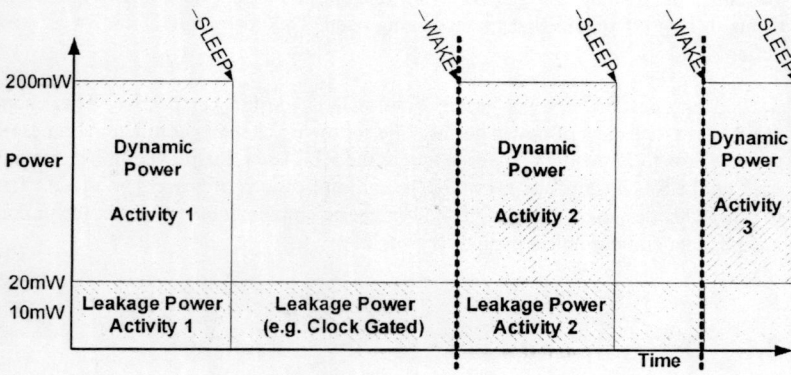

Figure 4-1   Activity Profile with No Power Gating

Figure 4-2 shows an example activity profile for the same sub-system with basic power gating implemented. The response time between the WAKE event and having clocks running may be significant and cannot be ignored at the system design level:

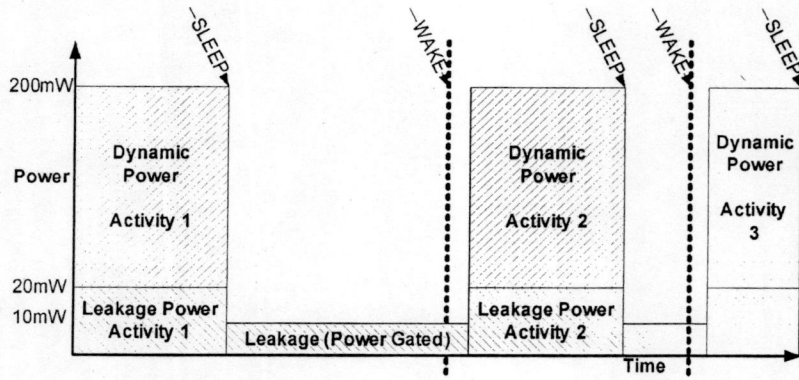

Figure 4-2   Activity Profile with Power Gating

Figure 4-3 shows that the leakage power savings are not perfect and instantaneous; the full leakage power savings take some time to reach target levels. This is due partly to the (hotter) thermal profile of the preceding activity and partly to the non-ideal nature of the power-gating technology. Therefore the achievable savings are compromised to some extent:

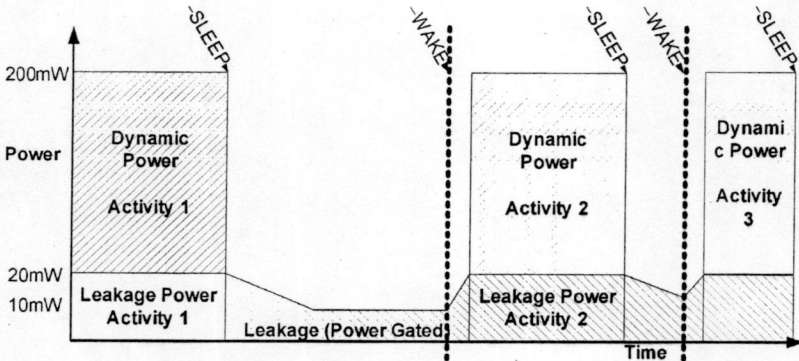

Figure 4-3   Realistic Profile with Power Gating

4.2   Impact of Power Gating on Classes of Sub-Systems

A cached CPU subsystem can typically be dormant or inactive for long periods, making power gating attractive. But there are some trade-offs that must be considered:

- Power gating the entire CPU provides very good leakage power reduction.
- But wake-up-time response to an interrupt has significant system level design implications (may even require deeper FIFO's or scheduled time-slots).
- If the cache contents are lost every time the CPU is powered down then there is likely to be a significant time and energy cost in all the bus activity to refill the cache when it is powered up.
- The net energy savings depend on the sleep/wake activity profile as to how much energy was saved when power gated versus the energy spent in reloading state.

A peripheral subsystem may have a much better defined profile than a CPU. It is under control of a device driver which can be profiled and operating system power management scheme which can be optimized. But there are still some trade-offs. In particular, it may be necessary to restore state quickly on wake-up to maximize power savings:

- The device driver may be required to explicitly load/restore key state or initiate hardware sequencer control as part of the sleep/wakeup sequence, but this places a significant burden on software.
- A better approach may be for the peripheral to store key state internally during sleep mode, but this requires special circuitry and additional control.

Finally, consider a more complex, multi-processor CPU cluster where one or more processors may be power gated off completely. In this case we assume that a processor is powered down only when it has completed a task and is idle, waiting for another task to be assigned. In this case:

- Power gating individual CPUs provides very good leakage power reduction.
- Because the CPU has completed its task, the fact that the local cache contents are lost when it is power gated is not a problem. The CPU is awoken clean and reset ready to execute and cache the next task it is given.
- Optimized energy savings may well require adaptive shutdown algorithms that vary the number of CPU cores power gated and active with varying workload.

In all these cases, power gating can provide significant leakage reduction in the design.

4.3   Principles of Power Gating Design

Power gating consists of selectively powering down certain blocks in the chip while keeping other blocks powered up. The goal of power gating is to minimize leakage current by temporarily switching power off to blocks that are not required in the current operating mode.

The most basic form of power gating control, and the one with the lowest long-term leakage power, is an externally switched power supply. Consider this example: an on-chip CPU has a dedicated off-chip power supply; that is, the supply provides power only to the CPU. We can then shut down this power supply and reduce the leakage in the CPU to essentially zero. This approach, though, also takes the longest time and requires the most energy to restore power to a gated block.

Internal power gating, where internal switches are used to control power to selected blocks, can be a better solution when powering down blocks for a short time.

Figure 4-4 shows a simplified view of an SoC that uses internal power gating.

Unlike a block that is always powered on, the power-gated block receives its power through a power-switching network. This network switches either V_{DD} or V_{SS} to the power gated block. In this example, V_{DD} is switched; V_{SS} is provided directly to the entire chip. The switching fabric typically consists of a large number of CMOS switches distributed around or within the power gated block.

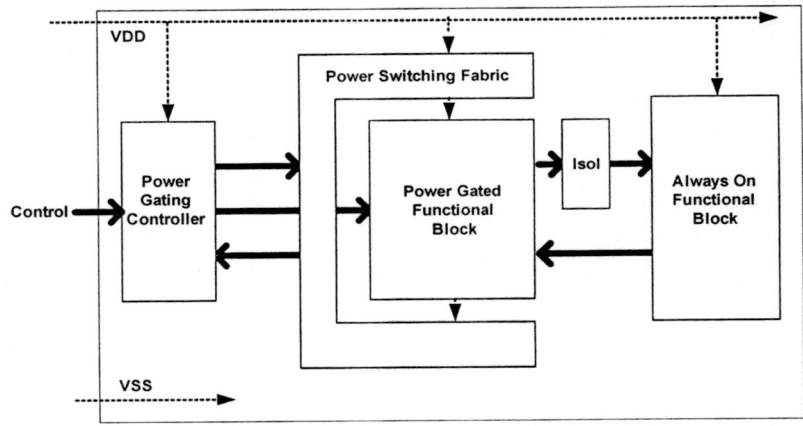

Figure 4-4   Block Diagram of an SoC with Power Gating

The power gating controller controls the CMOS switches that provide power to the gated block.

One challenge for power gating designs is that the outputs of the power gated block may ramp off very slowly. The result could be that these outputs spend a significant amount of time at threshold voltage, causing large crowbar currents in the always powered on block.

To prevent these crowbar currents, isolation cells (the "Isol" block in the figure) are placed between the outputs of the power gated block and the inputs of the always on block. These isolation cells are designed so that they do not experience crowbar current when one of the inputs is at threshold, as long as the control input is off. The power gating controller provides this isolation control signal.

For some power-gated blocks, it is highly desirable to retain the internal state of the block during power down, and to restore this state during power up. Such a retention strategy can save significant amounts of time and power during power up. One way of implementing such a retention strategy is to use retention registers in place of ordinary flip-flops.

Retention registers typically have an auxiliary or shadow register that is slower than the main register but which has much less leakage current. The shadow register is always powered up, and stores the contents of the main register during power gating. These retention registers need to be told when to store the current contents of the main register into the shadow register and when to restore the value back to the main register. This control is provided by the power gating controller.

4.3.1   Power Switching – Fine Grain vs. Coarse Grain

A critical decision in power gating is how to switch power. In general, there are two approaches: fine grain power gating and coarse grain power gating.

In fine grain power gating the switch is placed locally inside each standard cell in the library. Since this switch must supply the worst case current required by the cell, it has to be quite large in order not to impact performance. The area overhead of each cell is significant (often 2x-4x the size of the original cell). Figure 4-5 shows an example of a fine grain AND gate.

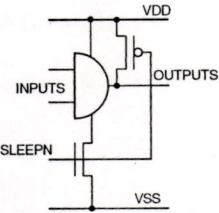

Figure 4-5   Fine Grain AND Gate with Pull-Up

The key advantage of fine grain power gating is that the timing impact of the IR drop across the switch and the behavior of the clamp are easy to characterize as they are contained within the cell. This means that it is still possible to use a traditional design flow to deploy fine grain power gating.

In coarse grain power gating, a block of gates has its power switched by a collection of switch cells. (Figure 4-4 on page 37). The sizing of a coarse grain switch network is more difficult than a fine grain switch as the exact switching activity of the logic it supplies is not known and can only be estimated. But coarse grain gating designs have significantly less area penalty than fine grain.

Over the last few years, there has been a strong convergence towards coarse grain power gating as the preferred method. The area penalty for fine grain power gating has just not proven worth the savings in design effort. Today, virtually all power gated designs use coarse grain power gating. For that reason, we focus exclusively on coarse grain power gating for the rest of this book. (The only exception is the chapter on libraries, where we will give a more detailed analysis of the fine-grain vs. coarse grain trade-offs.

One of the key challenges in any power gating design is managing the in-rush current when the power is reconnected. This in-rush current must be carefully controlled in order to avoid excessive IR drop in the power network; otherwise, the function and state of powered-on blocks could be corrupted as the power gated block goes through its sleep/wakeup sequence.

4.3.2   The Challenges of Power Gating

Implementing power gating presents certain challenges to the designer. These include:

- Design of the power switching fabric
- Design of the power gating controller
- Selection and use of retention registers and isolation cells

- Minimizing the impact of power gating on timing and area.
- The functional control of clocks and resets
- Interface isolation
- Developing the correct constraints for implementation and analysis
- Performing state-dependent verification for each supported power state
- Performing power state transition verification to ensure all legal state entry and exit arcs are simulated and verified
- Developing a strategy for manufacturing and production test

These topics are discussed in the following chapters.

CHAPTER 5   *Designing Power Gating*

This chapter describes power gating design from a front-end, RTL perspective. Figure 5-1 shows the critical components of such a design.

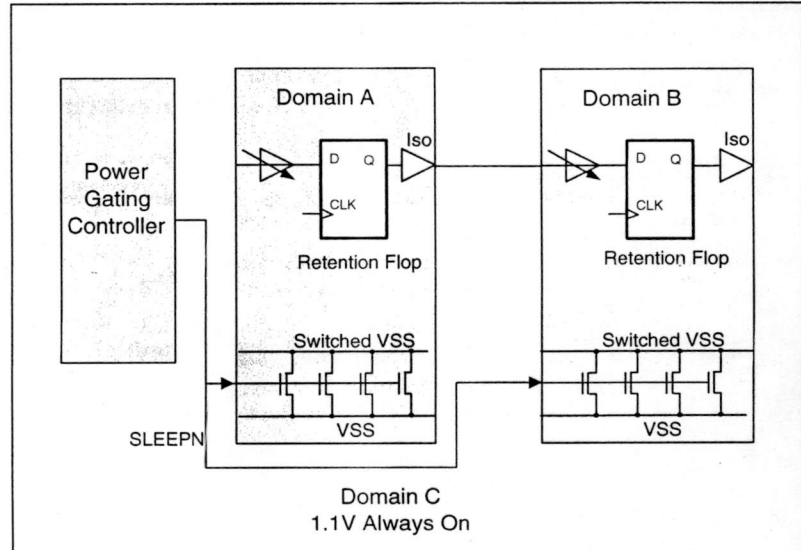

Figure 5-1   SoC with Power Gating

The critical issues in power gating include the design of the switching network and the power gating controller. We also need to determine when and where to insert retention flops and isolation cells.

5.1   Switching Fabric Design

The detailed transistor structures for power gating are highly technology specific and are described in detail in Appendix A. But we will consider here some of the architectural aspects of the switching fabric design.

The first architectural issue is whether to switch VDD (with a "header" switch) or to switch VSS (with a "footer" switch) or both.

A number of academic papers have been published on this subject. Some authors advocate both P-channel "Header" switches gating the VDD supply and N-channel "Footer" switches gating the VSS ground. However, two such high-V_T power switches in series with the gate cause a more significant IR voltage drop in the supply as seen by the gate. This drop in turn causes increased delays for the gates in the design.

In many practical designs this performance loss cannot be tolerated, and only one of the rails is switched.

Basic Header-Switch structure: Basic Footer-Switch structure:

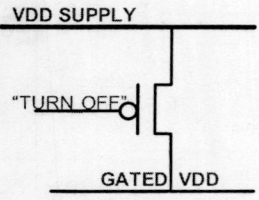
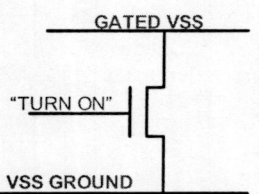

With a header-style switch fabric, the internal nodes and outputs of a power gated block collapse down towards the ground rail when the switch is turned off. With a footer-style switch fabric the internal nodes and outputs all charge towards the supply rail when the switch is turned off.

Note that here is no guarantee that the power gated nodes will ever fully discharge to ground or fully charge to the supply. Instead, an equilibrium is reached when the leak-

age current through the switches is balanced by the sub-threshold leakage of the switched cells. This is one of the reasons why isolation cells are required on outputs of power gated blocks, as discussed later in this chapter.

From a functional perspective, there are several arguments for using header cells:

External power gating (switching a power supply external to the chip) is only practical for switching VDD. VSS is usually common on the board for a variety of reasons, including providing a conduction path for ESD. If external and internal power gating are both used on the same chip, then switching VDD on chip will result in similar behavior under both power gated conditions; this simplifies functional verification, timing closure, and power analysis.

In SoC designs, the use of multiple power supplies is becoming increasingly common. These designs require level shifters on signals between blocks operating at different voltages. Level shifters are typically designed with a common ground and two different supply voltages. In chips using this design approach, switching the ground on power gated blocks can be a problem.

Finally, designers think in terms of "off" meaning signals are pulled to ground. It is just easier to think through all the system design issues when power, not ground, is switched.

The arguments in favor of the footer cell approach – that is, switching VSS – are based in the electrical characteristics of the switches themselves. This is discussed in Chapter 14.

Recommendations:

* Switch the supply rail or ground, rather than both, in order to minimize the IR drop.
* Decide early on in the design phase whether header or footer switches most naturally fit with the system design.
* Header switches may be the most appropriate choice for switches if external power gating will also be used on the chip.
* Header switches may be the most appropriate choice for switches if multiple power rails and/or voltage scaling will be used on the chip.

Pitfalls:

* Beware of mixing "footer" power-gating with externally switched power rails or multiple power supplies. This complicates functional, timing, and power analysis as well as placing more complex demands on the standard cell library.

5.1.1   Controlling the Switching Fabric

A key concern in controlling the switching fabric is to limit the in-rush current when power to the block is switched on. Excessive in-rush current can cause voltage spikes on the supply, possibly corrupting registers in the always-on blocks, as well as retention registers in the power gated block.

Various methods for controlling this in-rush current are described in Chapter 14. One representative approach is to daisy-chain the control signal to the switches. Each switching fabric typically will have hundreds (or more) switches acting in parallel; the control signal from the power controller is connected to the first switch, and it buffers (with an appropriate delay) the signal and sends it on to the next switch.

The result of this daisy chaining is that it takes some time from the assertion of a "power up" signal until the block is powered up. For this reason, switching fabrics will often provide an acknowledge signal indicating that the fabric is completely powered up. This signal can simply be the final buffered version of the "power up" control signal.

A more aggressive approach to turning on the switching fabric is to use several power-up control signals in sequence. The first control signal may turn on a set of weak or "trickle" switches, which initiate the power up but limit the in-rush current. The second control signal may then turn on the main set of power switches.

Regardless of the specific control method, during the power up sequence, it is important to wait until the switching fabric is completely powered up before enabling the power gated block to resume normal operation. The timing of this power up sequence is the responsibility of the power controller.

Note that the control signals for the power switching fabric – the whole daisy chain of power on/off and acknowledge – must be buffered by always on buffers, not by power gated buffers. This adds some level of complexity to the power routing of the power gated region.

5.1.2   Recommendations and Pitfalls for Power Gating Control

Recommendations:

- The power controller needs to be designed for the technology-specific power gating fabric used.
- Assertions should be provided for the power gating control ports, to match the chosen switch technology to ensure function verification and coverage in the RTL design environment.
- Power gating control signals must be made controllable during test.

Pitfalls:

- Combining external power gating (switching the external rail) with internal power gating (switching power on-chip) can be tricky. A supply that is "always on" during internal power gating may not be "always on" during external power gating. Careful design and verification is required in this case.

5.2   Signal Isolation

Once we have addressed the design and control of the switch fabric, the next problem is to determine the isolation strategy.

Every interface of a power gated region needs to be managed. We need to be sure that powering down the region will not result in crowbar current in any inputs of powered-up blocks. Also we need to be sure than none of the floating outputs of the power-down block will result in spurious behavior in the power-up blocks.

The outputs of the power gated block are the primary concern, since they can cause electrical or functional problems in other blocks. The inputs to the power gated blocks usually are not an issue – they can be driven to valid logic values by powered up blocks without creating electrical (or functional) problems in the powered down block.

5.2.1   Signal Isolation Techniques

The basic approach to controlling the outputs of powered down blocks is to use an isolation cell to clamp the output to a specific, legal value.

There are three basic types of isolation cell: those that clamp the signal to "0", those that clamp it to "1", and those that latch it to the most recent value.

In most cases, it is sufficient to clamp the output to an inactive state. When using active high logic, the most common approach is to clamp the value to "0". An AND-gate function accomplishes this. With active low logic, an OR-gate function parks the output at logic "1" .

Clamp library cells are designed to avoid crowbar currents and leakage paths when signal input floats, as long as the control input is in the appropriate ("isolate") state. In addition, their synthesis models typically have extra attributes to ensure these cells never get optimized away, buffered incorrectly or inverted as part of logic optimization.

The left side of Figure 5-2 shows a conceptual view of an AND-style isolation clamp-low. When the active low isolate signal "ISOLN" is high, the signal passes to the output; when "ISOLN" is low, the output is clamped low:

Figure 5-2   Basic Isolation Cells

The right side of Figure 5-2 shows a conceptual view of an OR-style isolation clamp-high. When the active high isolation control signal "ISOL" is high, the output is clamped high; when low the signal passes to the output:

These clamp gates add delay to the signals they are isolating. For some critical paths this added delay may not be acceptable – for example on cache memory interfaces.

An alternative isolation technique that does not add full gate delays is to use a pull-up or pull-down transistor. However this approach introduces multiple drivers on the power gated net, requiring careful sequencing to avoid contention. Even if the pull-up or pull-down transistors are relatively weak devices, the total number can be large enough that excess current from bus contention could cause problems. The sequencing to avoid contention is done by the power controller.

The left side of Figure 5-3 shows a conceptual view of a pull-down style clamp-low; when "ISOL" is high, the output is clamped low; when low the signal passes to the output:

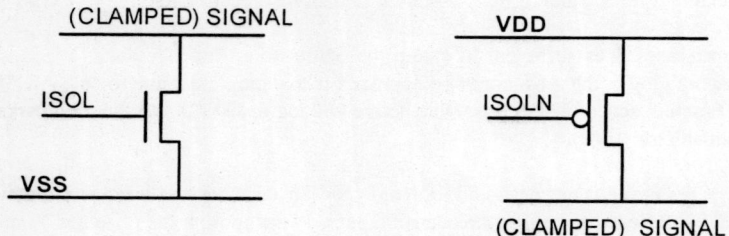

Figure 5-3   Pull-Down and Pull-Up Isolation Transistors

The right side of Figure 5-3 shows a conceptual view of a pull-up style clamp-high; when "ISOLN" is high, the signal passes to the output; when low the output is clamped high:

To avoid output glitches, it is important to isolate outputs during power up, and to keep them isolated until power has stabilized. This sequencing is straight-forward with the clamp cells but can be quite challenging with pull-up/pull-down transistors, since they would fight the output values whenever these powered back up in an active state.

Note that transistor-type clamps can cause metal migration and reliability problems when even a small amount of continuous current flows through them. They also create a big problem for test – any time there are multiple drivers on a net, testability becomes very difficult.

Therefore pull-up and pull-down clamps are not recommended for portable RTL design despite the lower area and timing cost. Instead, the "gate-style" cell styles are recommended, and described in the rest of this section. However, the pull transistor approach is useful in specialized situations where timing is critical, the signaling protocols are understood, and contention can be eliminated (and verified to be eliminated) by design.

5.2.2   Output or Input Isolation

As described above, it is necessary to isolate the outputs of a power gated block in order to avoid the electrical problem of floating outputs driving inputs of powered-up blocks. Logically, it makes no difference whether we clamp these signals at their source – that is, inside the power gated block – or at their destination – that is, in the powered-up blocks.

But there are important practical considerations that affect this choice.

It is likely that at least some of the outputs of the power gated block go to more than one powered-up block. If the outputs are isolated at the receiving blocks, more than one isolation cell may be required for each output. Therefore it is more area-efficient to isolate the outputs inside the power gated block.

Isolating outputs inside the power gating block also makes analysis easier. Once we have determined that all outputs are correctly isolated, we are done. If these signals are instead isolated in the receiving blocks, then each fan-out of the output signal must be checked to assure that there is an isolation cell on it. If the power gated block is reused in different applications, this analysis must be performed again in each situation.

Isolating outputs at their source does present some constraints for place and route, however. Unlike the other gates in the power gated block, the isolation cells must remain powered during power-down. Thus, the power domain containing the power gated block must now provide both switched power and always-on power. This somewhat complicates power routing of the chip, but modern EDA tools are capable of solving this problem.

Regardless of whether outputs are isolated at their source or their destination, the EDA tools must respect the unique character of isolation cells. If isolation is at the source, then the tools must not buffer the output of the isolation cell with power-gated buffers. If isolation is at the destination, then the tools must not buffer the (pre-isolation) signal with buffers that are always on. The control signal to the isolation cells must be buffered only by always on cells.

Also, optimization during place and route must not replace the isolation cell with a non-isolation cell.

For the reasons outlined above, we strongly recommend that reusable IP be designed with isolation cells within the IP, so that the complications of isolation are hidden from the SOC-level integration.

5.2.3   Interface Protocols and Isolation

In designing an interface of a power gated block there two goals: to minimize leakage and to avoid unnecessary or incorrect behavior.

Consider the case of a power-gated block that has outputs that go to an always on block. When the power gated block is powered down, the clamped signal values are received by the powered block. If these signals are active high, and they are clamped high, the destination may interpret these signals as commands, and act incorrectly. Clamping the signals to their inactive state is the best strategy for avoiding this problem. For most designs, this means clamping the outputs to "0".

The one possible exception to this guideline is reset. Typically, resets are active low, so that clamping it low signals a reset state on the interface. And in fact, this may be the most appropriate value to drive reset during power down. This assures that reset will be asserted during power up. (This is what we did on the SALT chip.) In any case, it is worthwhile to consider the effects of clamping interface signals to the active or inactive state during power down.

Consider now the case of a power-gated block that has outputs that go to another (independently) power-gated block. Because the blocks are independently power gated, their outputs must be isolated. But in some cases, the source block will be powered down and the destination block will also be powered down. In this case, clamp-

ing signals to the wrong level may increase leakage current. For example, if VDD is switched (for the destination block), and outputs (of the source block) are clamped to "1", then there may be sneak current paths from the clamped output to ground. This could cause unnecessary leakage.

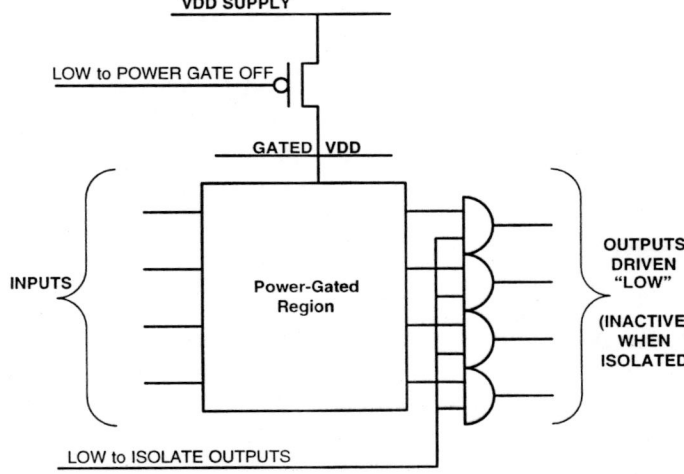

Figure 5-4   Output Isolation

Thus, for header-switched designs, the base-line recommendations are straight-forward: use active high signals and active low reset, and clamp all signals to "0".

For footer-switched power-gated regions we recommend active high signals and active low reset, and clamp all signals to "0", in general. There is one possible exception to this rule: if transmission gates are used at the inputs of the receiving block. If transmission gates are used, we may need to clamp to "1" to avoid sneak path leakage. This situation is discussed in detail in Chapter 12. In this case, it may be necessary to use active high signaling to avoid functional problems. Since most existing protocols use active high signals, this may present a design challenge.

Note: We have a general bias towards active high signals and active low reset for two reasons: this is the most commonly used design approach, and this is also leads to simple, easy-to-debug RTL.

A final note: for complex protocols, it may be necessary to use a more complex isolation strategy: clamping signals to their last value. This approach allows re-starting the protocol in process, instead of starting up in the reset state. This strategy requires a latched isolation cell, which is not available in all libraries, and so it should be used only if absolutely necessary.

5.2.4   Recommendations and Pitfalls for Isolation

Recommendations:

- Isolate the outputs of power gated blocks.
- Use isolation cells rather than pull-up or pull-down style clamps unless using very specialized interface protocols (where the "multiple-driver" challenges may be worth the implementation complications).
- Ensure that stuck-at-0 and stuck-at-1 faults can be detected during test on the isolation control signals. This facilitates verifying during manufacturing test that isolation works.

Pitfalls:

- Make sure the isolation cells really are always powered on.
- Isolation clamps on clocks can considerably complicate clock tree synthesis and timing closure. Clock tree balancing in particular can become difficult. If possible, avoid clocks that are generated in a power gated block and used externally to the block.

5.3   State Retention and Restoration Methods

Given a power switching fabric and an isolation strategy, it is possible to power gate a block of logic. But unless a retention strategy is employed, all state information is lost when the block is powered down. To resume its operation on power up, the block must either have its state restored from an external source or build up its state from the reset condition. In either case, the time and power required can be significant.

In many cases, an explicit retention strategy for saving and restoring state quickly and efficiently can provide a much faster and power-efficient method of getting the block fully functional after power up.

How essential a retention strategy is depends on the subsystem characteristics. A Digital Signal Processing unit that is primarily data-flow driven may be able to start from reset if it is supplied with new input data. However a peripheral or cached processor typically has enough residual state that the amount of bus traffic required to reload this state is excessive.

There are several methods for saving and restoring the internal state of a power gated block:

- A software approach based on reading and writing registers
- A scan-based approach based on using scan chains to store state off chip
- A register-based approach that uses retention registers

With the software approach, a processor in an always on block reads the registers of the power gated block during the power shutdown sequence. This state information is stored in the processor's memory. During the power up sequence, the processor reads its memory and writes the state back into the power gated block. This method has several drawbacks:

- The bus traffic considerably slows the power down and power up sequences.
- Bus conflicts can make the save/restore times non-deterministic, making it harder to decide when it is worthwhile shutting down the block.
- Software must be written and integrated into the system's software for handling power down and power up. This makes the software much less reusable and requires a much more extensive knowledge of the hardware on the part of the engineers writing the software.

For these reasons, we will focus on the other two approaches to retention, which make the power sequencing much more transparent to the rest of the system.

5.3.1   State Retention Using Scan Chains

Scan chains that are implemented for manufacturing test can be re-used to perform state retention with almost no incremental area overhead.

In this approach, a dedicated set of scan chains is used for the power gated block. During the power down sequence, the scan registers are shifted as in scan testing, but the outputs are routed to a memory. This memory can be on chip or off chip, but if on chip it needs to be always powered on. During the power up sequence the scan chains are loaded from the memory.

Note that once state is scanned out to memory, the entire subsystem can be power gated off. There is no need to keep an always on power region for retention registers.

From an RTL design perspective there are of course challenges. The most basic challenge is that scan flops are not inserted and connected up until synthesis – yet it is necessary to code and debug the controller at the RTL level, before synthesis.

Even the number of registers and the length of the scan chains are only known after initial implementation. Therefore the control sequencer needs to be parameterized to manage implementation-dependent counter values. It must also provide explicit control of scan enable and scan chains; these are later hooked up with the net-list.

To achieve the fastest save and restore times, we would like to write the retention data to the memory using the full width of the memory data bus. That implies that we should make the number of scan chains equal to the width of the memory data bus. In practice, this may be too many scan chains to be practical for manufacturing test. But typically we will use at least 8 scan chains, and potentially a multiple of 8.

If more that one chain is used then it is necessary to balance the scan chains – that is, they must all be the same length. This balancing is necessary because the controller generates a single shift enable signal that is shared by all the chains and is also used to gate data into and out of memory. To achieve this balance, we can add extra registers to the short chains.

Note that the retention memory must be large enough to hold the number of scanned bits. Also, there is a real-time delay cost in both saving and restoring state. This grows with the size of the block to be scanned out and back in, and is a function of how many scan chains are used.

There is also an energy cost in shifting the register state out and back in. If an external memory is used, then the IO switching power can be significant. Even using internal memory, there can be large dynamic power required just to shift the data through the scan chain. The patterns shifted are highly state dependent; in the pathological worst case, every flop in the block is toggling on every clock. This is much more toggling (and power) than the typical case, and can create an IR voltage drop that is unacceptable.

Modern test and implementation tools already need to deal with the fact that, during scan, toggle activity can be much higher than during normal operation. These tools can analyze the actual IR drop and allow the engineer to adjust the number of chains and the clocking sequence to keep the IR drop to an acceptable level. But care needs to be taken to avoid excessive IR drop that can corrupt data.

In spite of these challenges, scan-based retention can be useful in some situations. For long term sleep, the leakage savings achieved by completely shutting down an entire subsystem, especially by shutting down the external power supply, can be significant. This savings is even more significant if state can be restored through the scan chains rather than having start from the reset state after power up.

Figure 5-5 shows of scan-based save and restore, simplified to 4-bits to keep the drawing small. Note that one of the scan chains is shorter than the rest, so a flop is added to balance the chains. Once the scan chains are balanced the state can be saved to memory ("SCAN-OUT & SAVE STATE DATA") and later restored from memory

("SCAN-IN & RESTORE STATE DATA") such that every register has the original state.

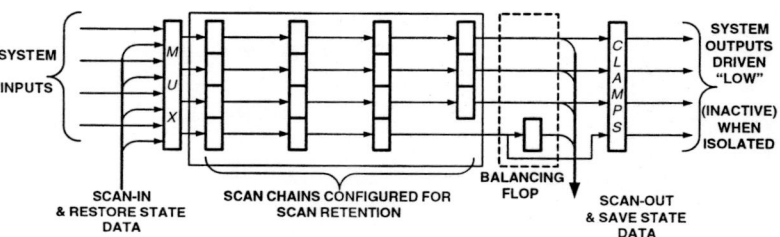

Figure 5-5   Scan Based Retention

Functional testing and simulation at the RTL level before netlist implementation (and scan insertion) is a challenge, but not insurmountable.

One approach is to add some conditional code into the RTL design which is only compiled when emulating scan-based retention. This code models the behavior of the shift registers and can be included in simple test sequences to verify that the controller is functioning correctly and the data is written to and read from memory correctly.

Below is an example of an RTL model of a dummy scan-chain for a CPU to be implemented with 16 scan chains for retention support. Note that dummy data is used for the scan chain.

```
`define CPU_SCAN_LEN 257 /* set to implementation
length once known */

`ifdef RTL_SLEEP_EMULATE
parameter scan_reg_length = `CPU_SCAN_LEN;
reg [15:0] scanword [0:scan_reg_length-1];
integer i;

/* initialize the scan chain to count pattern, or more
draconian X */
    initial
    begin
        for (i=0; i < scan_reg_length; i=i+1)
        begin
            scanword[i]<=i; // or 16'hXXXX;
```

```
    end
end

/* emulate scan shift CPUSI -> CPUSO */
always@(posedge CLK) begin
    if (CPUSE == 1'b1) /* when SCAN ENABLE is active */
    begin
        for (i=1; i < scan_reg_length; i=i+1)
        begin
            scanword[i]<=scanword[i-1];
        end
        scanword[0] <= CPUSI[15:0];
    end
end
assign CPUSO [15:0] = scanword[scan_reg_length-1];
`endif
```

At a later stage a gate level netlist simulation should be performed to ensure that the implementation-specific scan chains and control signals really are wired up correctly and that the correct length scan chain has been implemented and balanced.

5.3.2   Retention Registers

Another approach to providing state retention while power gating is to replace a standard register with a retention register. A retention register contains a "shadow" register that can preserve the registers state during power down and restore it at power up. Unlike the main register, the shadow register is always powered on.

Figure 5-6 shows two retention registers. In each case, the main register – the master and slave latches of the flop – is powered by the switched power rail "VDD_SW." The CLK, D, and RESETN pins all operate on the main register, which drives the Q output.

In addition, there is a shadow register "RET" which is used to save and restore state to the main register. The shadow register is powered by the always on rail "VDD."

With the register shown on the left side Figure 5-6, when SAVE is asserted, the state of the main register is loaded into the shadow register. When RESTORE is asserted, the state of the shadow register is loaded back into the main register. SAVE and RESTORE are level-sensitive signals.

With the register shown on the right side of Figure 5-6, when RETAIN goes high, the state of the main register is loaded into the shadow register. When RETAIN goes low,

the state of the shadow register is loaded back into the main register. RETAIN is an edge-sensitive signal.

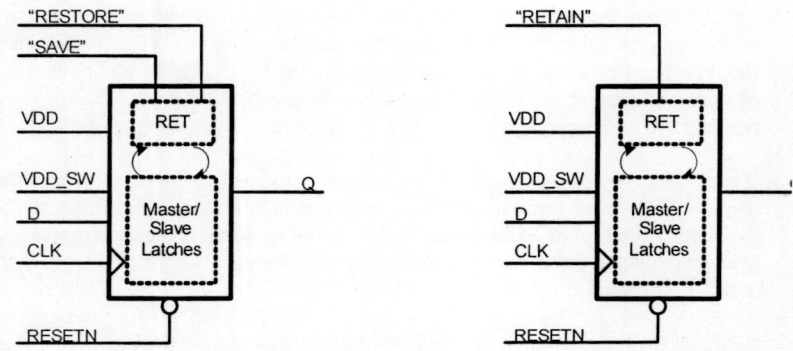

Figure 5-6   Retention Registers

The detailed design of retention registers is discussed in Chapter 13. We will just make a few comments about them here.

Real-world retention registers all have some area overhead, typically 20% or more. Some retention registers incorporate guard bands to isolate the retention state as robustly as possible from power gating transients. In this case the area overhead can be as large as 50% or more. In a design with a large number of registers this area impact can be significant.

Unfortunately, this overhead is unavoidable if an on-chip retention strategy is to be used. Consider the alternative. We could constrain the implementation to use only low leakage, high V_T registers, and connect these to the un-switched power rail. We then would simply power gate all the leaky (Low- or Mixed-V_T) combinatorial logic between register stages. However in any reasonable sized block the reset and clock networks typically have to be implemented with high-leakage low-V_T buffer trees to meet timing. These buffer trees contribute a significant portion of the leakage for the block, so they must be power gated. But as soon as these nets are power gated, the clocks and resets float and corrupt the registers.

In addition to the area penalty, the use of retention registers requires a more complex power controller.

5.3.3   Power Controller Design for Retention

The controller must manage the explicit sequencing of the save and restore signaling as part of the power management control state machine.

The shadow register may be quite slow compared to the main register, yet we need to make sure that the value in the main register is stable throughout the save operation. For this reason, most designers stop the clock before performing save. Of course, the save operation must be complete before power is shut down.

The restore cannot be performed until power is restored and the power gating transients have subsided. The restore operation must be completed before new values can be loaded into the main registers. For this reason, most designers do not re-start the clock until the restore operation is complete.

Although the power gating controller requires some care in design, the actual substitution of retention registers for standard flip-flops can be done automatically during implementation. Similarly, the connection of the save/restore control signals can be automatically connected to the retention registers. These control signals need to be implemented as always-on networks to avoid state corruption during power gating but otherwise can be treated transparently to the RTL design.

One of the advantages of the retention-register approach is that it allows the retention strategy to be largely transparent to the RTL designer. In this sense it closely follows the model of automatic scan insertion and hookup. To keep retention transparent to the RTL design, neither the clock nor reset can be active during retention. Otherwise, the RTL design would have to deal explicitly with conflicts between save/restore behavior and clock/reset behavior.

To minimize leakage, the clock and reset trees are likely to be off during power down. To keep these floating (X in simulation) signals from corrupting the retention register, retention must have priority over clock and reset. In designing the power gating controller, it is important to understand the behavior of the available cells in the power gating library to ensure that the shadow register is not corrupted due to floating clocks and resets.

5.3.4   Partial vs. Full State Retention

One of the key architectural decisions in power gating is how much state to retain during power down. Retaining the full state of the block – that is, replacing all registers with retention registers – provides the most robust, easily verified, and most transparent form of power gating.

In some designs, however, the area penalty for full state retention justifies considering partial state retention - retaining only some of the internal state of the block. But partial state retention poses some significant challenges.

In partial state retention, only the "architecturally visible" state is saved and restored. The challenge is to assure that all non-retained registers power up in legal, safe and verifiable states.

Example candidates for non-retained registers are FIFOs, memories, and counters. Converting these devices to retention registers can be quite expensive. In some designs, it may be appropriate just to make sure that they power up to a known state. For FIFOs, this would mean resetting the FIFO controller to indicate that the FIFO is empty. For the memories, we could reset the memory controller so that it considers the memory uninitialized. We could reset the counters to zero.

The concept of deep and shallow state might be useful in determining candidates for non-retained registers. Shallow state refers to the registers that directly control the logic of the design – that is, the part of the design that could be drawn as a state machine diagram. Deep state refers to registers that are used by the state machine but which contain large amounts of auxiliary data – like memory, counters and FIFOs. We would not normally draw these registers as part of a state machine diagram.

For partial retention, a reasonable strategy may be to save and restore shallow state, and to have a separate strategy for dealing with deep state. Resetting the controllers of the deep state registers is one possible strategy.

The problem then becomes how to verify that on power up, the combination of retained and non-retained state allows the block to restart correctly. During simulation, we set the outputs of all the registers (both deep and shallow state) to X. That is, we corrupt all registers except the shadow registers in the retention flops.

The key to verifying correct start-up after power gating is to assure that the X's are not propagated. That is, after the power up sequence is complete, there should be no X's in the circuit except the contents of memory. And the X's in memory should not be able to propagate and affect the function of the circuit.

The careful and selective use of resets can solve this problem. But we need to make sure that we only reset non-retained registers, so that we do not interfere with the restore function of the retention registers.

Thus, it becomes important to have separate reset signals for retention and non-retention storage. We can then architect the power up sequence after power gating to restore retained state and initialize all the non retention registers.

The power controller for partial retention must therefore drive independent (named) resets to the appropriate portions of the subsystem. Some rigorous functional testing is required to ensure that there are no illegal combinations of states that might cause deadlock.

5.3.5   System Level Issues and Retention

A more subtle complication arises from a potential interaction with clock gating, which is implemented further down the design flow. All the state bits that make up enable terms for clock gating need to be carefully managed: either retained or be re-initialized to a safe and restart-able condition. In this way, the contents of the transparent latch in the clock gating cells can cleanly be regenerated – without the requirement to add retention to the clock gating cells.

Similarly, using both edges of the clock can be a real problem. In the power down sequence, the clock is stopped in the "0" state. This leaves the clock gating latch transparent; when power and state are restored, the terms that form the clock gating control propagate through the latch, restoring the correct value to the inputs of the clock gated registers.

If both positive-edge-clocked flops and negative-edge-clocked flops are used in the same design, then there is no value that we can park the clock that will leave all the clock gating latches transparent. Thus, we will not be able to restore all the data correctly.

Retention also makes scan testing somewhat more complicated. In order to perform scan testing, we need to force the retention flops into their normal operating mode. Thus, when we enter scan mode (for the power gated block) we need to set the power controller so that save and restore are both de-asserted. When we enter scan mode for the power controller itself, we need to relax this constraint, so that scan can be used to test the generation of the save and restore signals.

5.3.6   Recommendations and Pitfalls for state retention

Recommendations:

- If partial retention is implemented then provide separate resets for the retained and the non-retained storage portions of the design. This allows clean verification of power on reset and restore/re-initialize operation.

- When implementing partial retention ensure that state machines and sequencers have no dependencies on non-retained state, in order to avoid state-dependent deadlock or invalid state conditions. (The state space to verify can be enormous if many retention state values must be tested with non-retained state).

- Where the area impact of specialized retention registers is too high then reusing the manufacturing scan chains is an option. Although this requires some care to map cleanly onto the netlist implementation after test structures have been generated, this can be managed relatively cleanly in RTL-coded control state machines.

- Retention controls must be made controllable and observable during scan test.

Pitfalls:

- Poor in-rush current management or retention power supply noise has the potential to corrupt retention registers resulting in unsafe/invalid state on restart. Great care must be taken in the RTL power control (and in physical implementation) to ensure power is reapplied safely.

- Partial retention requires much more rigorous reset and restore validation to ensure there are never deadlock conditions between retained/restored state and re-initialized non-retained state.

- Clock gating enable terms that affect retention state need to have retention registers on their entire fan-in state in order to ensure that "next state" sequencing behaves correctly.

- A scan-based save and restore approach is likely to use the system bus to transfer data to/from memory. This bus typically can have wait states; thus, care needs to be taken to ensure the scan save/restore controller can support wait-states without any data loss.

- During manufacturing test, failure to test that retention registers actually retain data can lead to failures in the field.

5.4   Power Gating Control

Given a power switching fabric, and isolation strategy, and a retention strategy, we can now design the power controller that controls the power down and power up sequencing.

5.4.1   Power Control Sequencing

From the discussions above, we extract the following requirements for power gating a block without retention.

To power gate a region without retention:

- Flush through any bus or external operations in progress
- Stop the clocks, in the appropriate phase to minimize leakage into the power-gated region
- Assert the isolation control signal to park all outputs in a safe condition
- Assert reset to the block, so that it powers up in the reset condition
- Assert the power gating control signal to power down the block

To restore power:

- De-assert the power gating control signal to power back up the block
- Optionally sequence multiple control signals for phased power-up depending on the current in-rush management approach and technology
- De-assert reset to ensure clean initialization following the gated power-up
- De-assert the isolation control signal to restore all outputs
- Restart the clocks, without glitches and without violating minimum pulse width design constraints

Figure 5-7 shows the power control sequencing for a power gated block without retention.

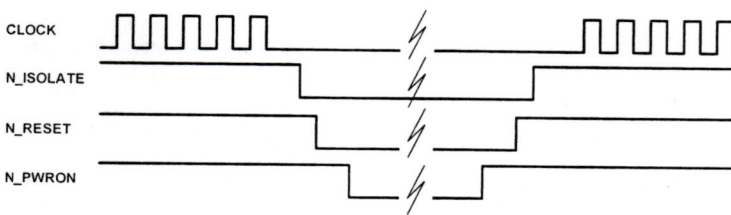

Figure 5-7   Power Control Sequencing Without Retention

For a block with retention, we must add the save and restore signaling to the power gating sequence.

To power gate a region with retention:

- Flush through any bus or external operations in progress
- Stop the clocks, in the appropriate phase to minimize leakage into the power-gated region
- Assert the isolation control signal to park all outputs in a safe condition
- Assert the state retention save condition (pulse or edge-triggered depending on the technology)
- Assert reset to the non-retained registers in the block, so that they power up in the reset condition
- Assert the power gating control signal to power down the block

To restore power and retained state:

- De-assert the power gating control signal to power back up the block
- Optionally sequence multiple control signals for phased power-up depending on the current in-rush management approach and technology
- De-assert reset to ensure clean initialization following the gated power-up
- Assert the state retention restore condition (pulse or edge-triggered is technology dependent)
- De-assert the isolation control signal to restore all outputs
- Restart the clocks, without glitches or violating minimum pulse width design constraints

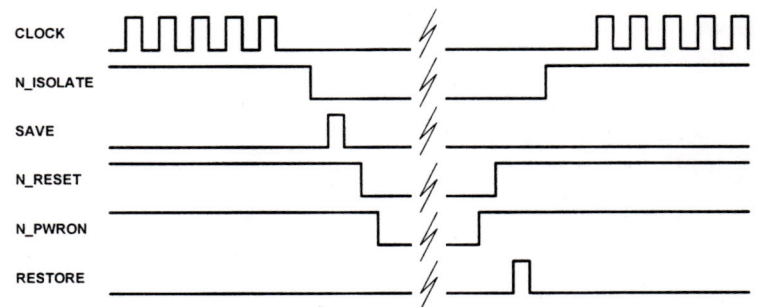

Figure 5-8   Power Control Sequencing With Retention

5.4.2   Handshake Protocols

Power gating takes time. The power gating switch fabric must be designed to limit voltage spikes that might corrupt retention registers or other powered-up logic. Most designs achieve this by limiting the current during power up, and thus limiting the rate at which the voltage rises to its final value.

The power controller must accommodate this process. In particular, it must wait until power up is complete before issuing restore. That is, it must insert a delay between power on and restore.

The simplest way to do this is to build a fixed delay into the controller sequencing. Counters can add in enough clock cycles to meet the power up or power down times. However embedding such time constants in the RTL ties the RTL to the timing of a particular switch fabric implementation. The result is that the IP is significantly less portable or reusable. Even migrating a working product onto a next generation technology node, where the power gating timing would be different, would require changes to the RTL.

For this reason, we recommend using a request-acknowledge handshake to control the power switching fabric.

Figure 5-9 shows an example of this protocol. The power controller issues a N_PWR_REQ to turn the power switching fabric off. It is the responsibility of the switching fabric to return N_PWR_ACK when power is completely switched off. On power up, the controller de-asserts N_PWR_REQ to turn the switching fabric on. When the fabric is completely on and it is safe to proceed, the switching fabric de-asserts N_PWR_ACK. When the controller sees the acknowledge, it proceeds to assert restore and continue through the power up sequence.

Chapter 14 provides more detail on how the switching fabric can be designed to provide the acknowledge signal.

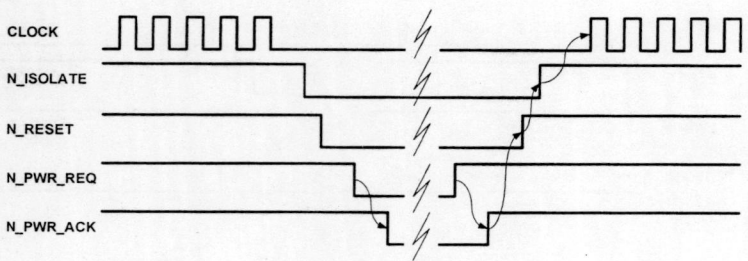

Figure 5-9   Power Switching With Acknowledge

In many applications, the power acknowledge signal is asynchronous – its timing depends on the switching fabric design. For this reason, the power controller needs to synchronize the acknowledge to its own clock before using it.

5.4.3   Recommendations and Pitfalls for power gating controllers

Recommendations:

- Design the control sequencers with request and acknowledge handshakes for the power gating control.
- Build in interlocks and synchronization to ensure a safe wake-up sequence.

Pitfalls:

- One critical case is when the controller tries to power the block up immediately after power down, and when in fact the power down is not complete. That is, the power up sequence starts while the power gating fabric is only partially powered down. Designers need to consider this case carefully in designing the power controller. Remember that the power down time is dependent on semiconductor process and temperature.

5.5   Power gating design verification – RTL simulation

We next consider the issue of verifying a power gated circuit at the RTL level. This is a challenge because Hardware Description Languages (HDLs) do not provide a mechanism for describing power connections at the RTL level. To simulate power gating we need to extend Verilog – either by modifying the code or by using a separate set of commands to describe power connections and power switching.

The Unified Power Format (UPF) defines both a language format and simulation semantics for power gating. Much of the UPF standard addresses the implementation of power strategies; this aspect is discussed in Chapter 11: Implementing Power Gating. Here we will limit our discussion the issue of simulating power gating.

EDA companies are moving rapidly to implement UPF and to provide the ability to simulate power gating automatically. For those who are using simulators that do not yet support UPF, it is possible to implement much of the UPF simulation semantics by adding special code to the RTL, either manually or by means of a script.

(Note: The script-based approach we describe here assumes a rigorous RTL coding style, such as that described in the *Reuse Methodology Manual*. It also depends on using a consistent naming scheme for clocks and resets.)

The key capabilities needed to simulate power gating at the RTL level include:

- functional modeling of power gating (including forcing outputs to X when power gated)
- functional modeling of isolation
- functional modeling of save and restore
- functional modeling of the precedence of power gating/retention/reset

In addition to simulation, assertions and functional coverage should be added in order to validate the correct sequencing and polarity of the control networks.

5.5.1 Inferring Power Gating Behavior in RTL

The first step is to simulate the effects of powering down a block.

UPF provides a mechanism (a set of tcl commands) for defining a power domain (a set of Verilog modules) and a set power supplies (power and ground supply nets) to the power domain.

Figure 5-10 on page 65 shows the power connections for the design we want to simulate. The Verilog module *my_module* (instance U1) has a header switch that controls power to all the logic in the module. The power gating controller de-asserts *pwr_req* to power down the module and asserts *pwr_req* to power the module up. The signal *pwr_ack* is the acknowledge signal that indicates that the switch has completed its power up/power down. At the RTL level it is a just a buffered version of *pwr_req*. In the gate level netlist, it will have real delays.

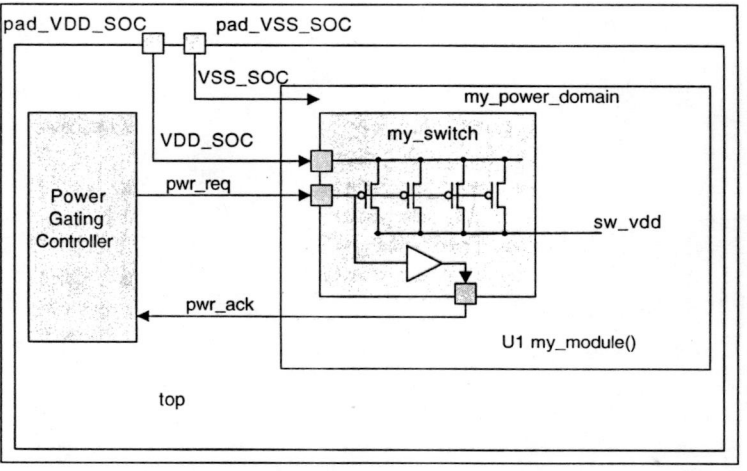

Figure 5-10

The UPF for such a design might look like this:

```
create_power_domain top -include_scope

create_power_domain top_power_domain -include_scope

create_supply_net VDD_SOC  -domain top_power_domain
connect_supply_net VDD_SOC -ports {pad_VDD_SOC}

create_supply_net VSS_SOC  -domain top_power_domain
connect_supply_net VSS_SOC -ports {pad_VSS_SOC}


set_scope U1
create_power_domain my_power_domain -include_scope

create_supply_net sw_vdd -domain my_power_domain

set_domain_supply_net my_power_domain
    -primary_power_net sw_vdd
    -primary_ground_net /top/VSS_SOC
```

```
create_power_switch my_power_switch
    -domain my_power_domain
    -input_supply_port {my_sw_input_port /top/VDD_SOC}
    -output_supply_port {my_sw_output_port sw_vdd}
    -control_port {my_sw_control_port  /top/pwr_req}
    -ack_port {my_ack_port /top/pwr_ack}
    -on_state {pwr_on_state my_input_port
                {my_sw_control_port ==1 })}
    -off_state {pwr_off_state
                {my_sw_control_port == 0}}
```

If we include this UPF code when we simulate, then the following will happen:

- When *pwr_req* goes low (requesting a power down), then the switch turns power off to all the elements in U1. That is, all the registers have their outputs set to X and all the output ports of U1 are set to X. All processes become inactive. At the same time, *pwr_ack* is set low, informing the power controller that the power is completely turned off. UPF supports assigning a delay for the acknowledge signal, but for RTL simulation we use the default of zero delay.

- When *pwr_req* goes high (requesting power to be restored), then the switch turns power on to all the elements in U1. That is, all the registers resume their normal operation and all the continuous assignment and combinational processes resume. At the same time, *pwr_ack* is set high, informing the power controller that the power is completely turned on.

If we are using a simulator that does not support UPF, we can accomplish similar behavior by modifying the RTL directly.

To do this we need a consistent asynchronous reset (or set) and synchronous clocking style to all sequential statements in the RTL. It is then possible to script a conditional set of power gating and behaviors that allow rigorous simulation modeling:

- Force "X" on all register outputs when power-gated
- Ensure that internal state is set to "X" when power gated to verify that reset actually resets state after power gating
- Model correctly the priorities of power gating/reset/clocking to ensure correct sequencing

For example, consider the following code:

```
always @ (posedge clk or negedge nrst) begin
    if (!nrst)
        current_state  <= 4'b0101;
    else
        current_state <= next_state;
end
```

This can be automatically converted to code of the form:

```
`ifdef  RTL_PG_EMULATE
    wire sw_vdd;
    assign sw_vdd  = pwr_req & pwr_ack;
`endif

always @ (posedge clk or negedge nrst
`ifdef  RTL_PG_EMULATE
     or negedge sw_vdd
`endif
) begin
`ifdef  RTL_PG_EMULATE
    if (!sw_vdd)
        current_state  <= 4'bXXXX;
    else
`endif
    if (!nrst)
        current_state  <= 4'b0101;
    else
        current_state <= next_state;
    end
end
```

When simulating with *RTL_PG_EMULATE* defined, the power gating signal *sw_vdd* is added to the sensitivity list of the process. It is the highest priority term in the sequential process description and forces *current_state* to X whenever power is removed.

Thus, we can write a script to modify every sequential process in *my_module* to have the power gating behavior coded above. Note that additional code must be added at the top level to connect *pwr_req* and *pwr_ack* to the power gating controller.

Also note that the behavior simulated here is not identical to the UPF semantics – only the register outputs are set to X, not the output ports of the module. So if any output is purely a combinational function of inputs, then that output will not be set to X by the modifications outlined above. Additional scripting is required to set such an output to X.

5.5.2 Inferring Power Gating and Retention Behavior in RTL

For designs that use retention, the next step is to modify the RTL to model the retention behavior:

- Initialize the retention state variable to "X" to capture invalid RESTORE before SAVE operation
- Sample register state to an extra inferred retention state variable for "SAVE" operation
- Force "X" on all register outputs when power-gated
- Re-initialize state from the retention state variable on "RESTORE" operation
- Model correctly the priorities of power gating/retention/reset/clocking to ensure correct sequencing

Again, we can do this either by adding to our UPF code or by writing a script to modify the RTL directly.

UPF provides commands to specify the "always on" power net for the retention registers and the save and restore control signals. By default, the ***set_retention*** command will convert all registers in the power domain to retention registers.

```
set_retention my_retention_strategy
    -domain my_power_domain
    -retention_power_net VDD_SOC

set_retention_control my_retention_strategy
    -domain my_power_domain
    -save_signal   {SAVE posedge}
    -restore_signal   {NRESTORE negedge}
```

UPF specifies the following semantics for these commands. We start with the same example as before:

```
always @ (posedge clk or negedge nrst) begin
    if (!nrst)
        current_state   <= 4'b0101;
    else
        current_state <= next_state;
end
```

With the added UPF code, the simulator will behave as if we had added the following two processes to the RTL:

```
reg [3:0]  save_current_state;
always @ (posedge SAVE) begin
    save_current_state <= current_state;
end
always @ (negedge NRESTORE) begin
    current_state <= save_current_state;
end
```

Note that this approach implies that *NRESTORE*, *clk* and *nrst* must be mutually exclusive; only one can be active at any time. Otherwise a conflict can arise between the different processes driving *current_state*. The power gating controller must be designed to comply with this requirement. We can use assertions to check during simulation that this requirement is not violated.

If we are not using a simulator that supports UPF, we can write a script that makes these same modifications to the RTL.

The resulting code would be added to the RTL:

```
`ifdef  RTL_PG_EMULATE
    reg [3:0]  save_current_state;
always @ (posedge SAVE) begin
    save_current_state <= current_state;
end
always @ (negedge NRESTORE) begin
    current_state <= save_current_state;
end
`endif
```

Note that with UPF we can control whether we simulate retention simply by whether we include the UPF in the list of source files. When we modify the RTL using a script, we need to use an *ifdef* to control simulation.

5.6 Design For Test Considerations

A key component of design for test (DFT) best-practices is providing external control of clocks and resets. This enables standard Automatic Test Pattern Generation tools to generate high coverage test vectors.

Power gating designs create some additional challenges for design for test. These challenges include:

- External control and observability for the power gating, retention and isolation signals
- Dealing with max current and power limitations during test
- Testing the power switching network for correct analog behavior
- Testing correct shutdown, isolation, retention behavior
- Testing correct function of power gating controller

5.6.1 Power Gating Controls

Best practice RTL design requires the designer to ensure controllability of resets for testability. All derived or re-synchronized resets (or presets) are multiplexed from an externally controllable primary reset control pin. That is, in test mode we must be able to override all the resets in the system and provide a master reset from an external pin.

For similar reasons the designer needs to provide controllability of power gating control networks. During test, we need to be able to:

- Prevent scan test patterns from accidentally toggling state machine outputs that activate power gating of sub-systems.
- Prevent scan test patterns from accidentally toggling isolation clamp signals.
- Prevent scan test patterns from accidentally asserting restore and corrupting the data in the scan flops.

Thus, all the signals coming out of the power gating controller need to be gated or multiplexed when in test mode. Forcing the isolation signals and restore off during scan is a minimum requirement. A better solution is to provide direct control over these signals from external pins or an on-chip test controller when in test mode.

Forcing power gating off – forcing all power gated block into power-up mode – during test is an option in some designs. But in many designs this is not an acceptable practice because of overall power limitations on the chip.

5.6.2 Power Limitations During Scan Test

During scan testing, all the flops in the scan chain can potentially toggle in each clock. This means that switching activity (and hence dynamic power) can be much higher during test than in normal operation. In fact, the dynamic power during scan test can exceed the capabilities of the package, leading to excessive heat and damage to the chip.

For this reason, we would like to be able to power down all the power-gated blocks in the chip except for the one under test. To do this, we need to be able to control the power gating signals from external pins during test mode. We also need to design the scan chains so that there are separate chains for each power gated block. We cannot have the scan chain for block under test go through a block that is powered down.

Depending on the design, it may be possible to multiplex the IO pins of the chip to provide the required control. In other designs, we may need to use a JTAG controller and some dedicated logic to control these power gating, isolation, and retention signals.

5.6.3 Testing the Switching Network

Manufacturing problems in the switching network are difficult to detect. Control buffer or switch transistor faults may lead to some power gates not being switched on properly, resulting in excessive IR drop. This can lead to the end product not meeting its performance specification.

Other defects may cause some power switches to be permanently on, resulting in excessive current consumption. This situation can be partially tested using IDDQ test, but may not always be detectable. IDDQ threshold(s) should be set to validate any needed specifications for battery life.

At-speed testing is an automated method that is able to identify some malfunctioning power switches. High impedance or broken power switches may cause timing failures in critical paths. Transition fault testing will pick up many of these, and targeted path delay testing can address others. No solution is fool-proof, however, so some functional test may be needed. Care must be taken during test development to ensure that all necessary clock control is available for each power mode under test.

Some form of static current (IDDQ) testing is needed to verify that power switches turn off correctly. For power-gated chips, the chip is put into a number of quiescent states (at least one for every "sleep" mode), and IDDQ is measured and compared with its specified value. Leakage measurements can take a long time, but specialized measurement techniques and DFT can both reduce test time and improve the quality of results. Because background leakage can be high, modern IDDQ testing methods often compare multiple measurements, sometimes across multiple die, rather than setting a single threshold value.

In addition to these tests, it may be useful to confirm that power domains can be powered up and down without corrupting the behavior and register contents of other (powered-up) blocks. In addition, we need to test that retention registers retain state when other blocks power up and down. Such tests can be quite challenging, since the effects of powering up a block can be highly design-dependent, as well as dependent on how much activity is going on in the adjacent blocks. But in some designs it may be worthwhile to develop functional vectors to verify this behavior.

5.6.4   Testing Isolation and Retention

During normal scan testing, we force the isolation control signals to the non-clamping state. The correct function of the isolation cells (in the non-clamped state) is then tested as part of the normal scan testing of the chip.

We can test the isolation cells in the clamped state in two different ways:

- We can use functional tests
- We can repeat the scan tests of the receiving blocks while clamping the isolated outputs of the block under test. With the isolated outputs clamped to a known value, these just become fixed inputs to the other blocks of the chip.

Manufacturing tests of retention registers require that both zeroes and ones can be saved and restored. This can be achieved by a special scan test where:

- a pattern of alternating ones and zeros are scanned into the flops
- save is asserted (from our external control)
- the complementary pattern of alternating ones and zeros is scanned into the flops
- Optionally - the block is powered down and then powered up
- restore is asserted
- scan out the results and check that the flops were restored correctly

We can then repeat the test with the reverse patterns of ones and zeros.

Note that if we include power down/power up in the test, the tester must be able to control the power gating of the block under test.

5.6.5   Testing the Power Gating Controller

With the techniques described above, we can provide effective manufacturing tests for power gated designs, with one exception: we have been forcing the outputs of the power gating controller. Now we need to test it.

We can test the power gating controller either with functional tests or with scan. The functional tests will be design specific. The scan test approach requires that we force the outputs of the power controller outputs to the appropriate state during scan – so that we avoid accidentally toggling power meshes up and down during the test. One way to facilitate this control is to wrap the power gating controller in an IEEE 1500 wrapper. This approach ensures full controllability and observability of the power gating controller while allowing us to keep the outputs of the controller at a stable value.

Recommendations:

- Clock and reset signals must be made externally controllable during test
- Power gating control signals must also be made externally controllable during test
- Isolation control signals need to be made controllable during scan test
- Retention controls must be made controllable during scan test.
- Support for IDDQ testing should be provided in the case where "stuck-on" power gates could potentially cause product malfunction in the end-customer system.

Pitfalls:

- Determining the appropriate target values for IDDQ testing is a challenge. Power gated quiescent current measurements can only be relative to full-on current measurements due to the wide spread in leakage currents across fabrication process. One approach is to require each measurement to have a value specified relative to the others, such as requiring the sleep state leakage to be less than 20% of operating leakage, for example.

CHAPTER 6

Architectural Issues for Power Gating

This chapter discusses some of the architectural issues involved in implanting power gating designs. In particular, it addresses the issues of partitioning, hierarchy, and multiple power-gated domains.

6.1   Hierarchy and Power Gating

A scalable approach to chip architecture is valuable since a system-on-chip design today often becomes a component in an even larger chip in a subsequent product generation.

To support this portability, module boundaries must be enforced at the power domain level. That is, a given module should belong to a single power domain, not split across several domains. Some tools and flows support RTL process by RTL process assignment to power domains, but this leads to much more complicated implementation and analysis. Clean visibility of the boundaries of a power-gated block is key to having a clean, top-down implementation and verification flow.

Although one can in theory nest power gated modules arbitrarily within power gated subsystems which are in turn nested on a shared switched power rail, there are considerable benefits in not creating multiple levels of power switching fabric. As described in Chapter 11, power-gating is intrusive and adds in some voltage drop and degradation of performance. Cascading multiple voltage drops can lead to unacceptable increases in delay.

Even if the design is represented as hierarchical at the architectural level, the implementation is improved if this is mapped onto a single level of power gating at implementation. Consider the example shown in Figure 6-1. The CPU conceptually has all the core logic power gated, and within it a number of functional units that can each be powered down independently – a Multiply-Accumulate and a Vector Floating Point units in this case:

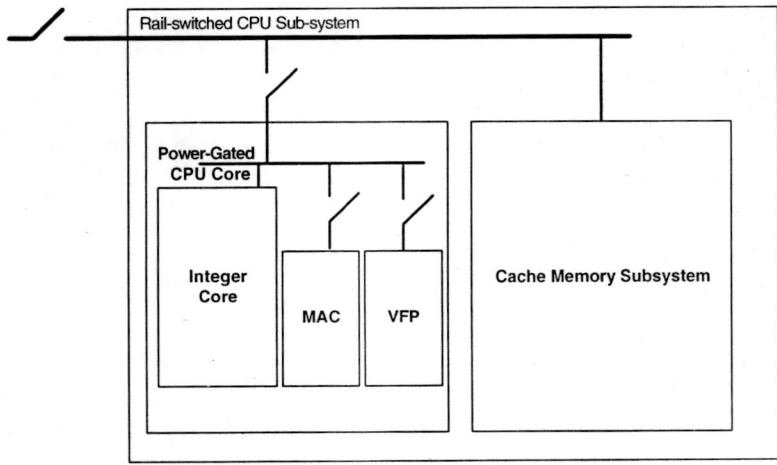

Figure 6-1   Power Gating Example

The modes of operation can be described in table form as:

Cache	CPU	MAC	VFP	Power State
(OFF)	(OFF)	-	-	Shutdown (Cache cleaned, VDDCPU off)
ON	OFF	-	-	Deep Sleep (Cache preserved)
ON	ON	OFF	OFF	Normal Operation
ON	ON	ON	OFF	DSP workload
ON	ON	OFF	ON	Graphics workload
ON	ON	ON	ON	Intensive multimedia mode

From an implementation standpoint the switching fabric is flattened as shown in Figure 6-2. There is never a case when the MAC or VFP functional units is switched on without the CPU core also being powered. So the switch control semantics are adjusted to AND the control terms rather than cascade the switch elements.

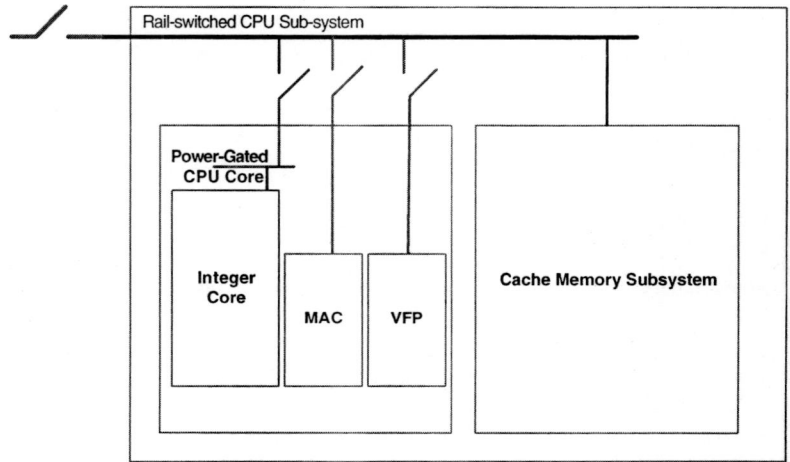

Figure 6-2   Flattened Switching Network

The power mode table now includes explicit control of the nested power gated functional units:

Cache	CPU	MAC	VFP	Power State
(OFF)	(OFF)	(OFF)	(OFF)	Shutdown (Cache cleaned, VDDCPU off)
ON	OFF	OFF	OFF	Deep Sleep (Cache preserved)
ON	ON	OFF	OFF	Normal Operation
ON	ON	ON	OFF	DSP workload
ON	ON	OFF	ON	Graphics workload
ON	ON	ON	ON	Intensive multimedia mode

Recommendations:

- Map power gated regions to explicit module boundaries
- When partitioning a hierarchical power gating design ensure that the power gating control terms can be mapped back to a flat switching fabric.

Pitfalls:

- Avoid control signals passing though power-gated or power-down regions to other power regions that are not hierarchically switched with the first region.
- Avoid excessively fine power gating granularity unless absolutely required for aggressive leakage power management. Every interface adds implementation and

verification challenges and complicates the system level production test challenges.

- Avoid a power gating system of more than one or two levels.

6.2   Power Networks and Their Control

In the design of a processor-based SOC the CPU system may well introduce a number of power networks:

- An independent power rail to the entire cached CPU subsystem – this allows the CPU to be completely turned off for long-term "sleep" modes of operation.
- A power gated supply to the CPU logic to support short-term leakage savings modes where the cache memory can be left retained but all the leaky standard cell logic turned off locally.
- Optionally, some form of always-on retention power supply from the non power gated rail. This is needed to support state-retention registers in the standard cell portion of the design.
- An always-on supply to provide power to the isolation cells.
- A non-power gated supply for the power gating controller and for the buffers on all the power control signals: the power switch controls, the retention controls, and the isolation controls.
- An SOC-level always-on supply to control the external rail switching handshake with the power supply.

Figure 6-3 illustrates the power networks with independent "VDDCPU" and always-on "VDDSOC" with a common VSS ground connection; in this example the power-gated standard cell area has a non-gated state retention supply shown to indicate an active supply rail within a power gated region:

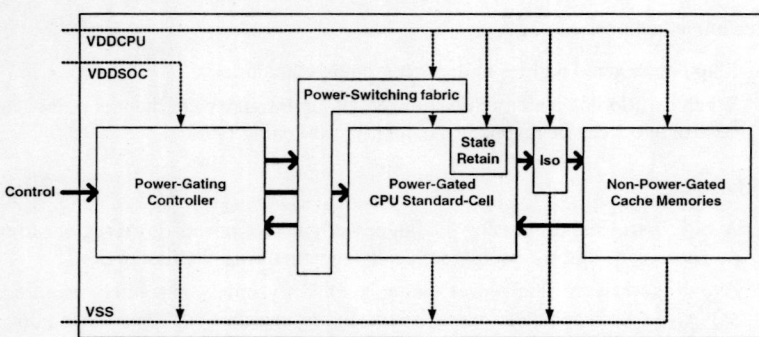

Figure 6-3   Power Network Control

6.2.1   External Power Rail Switching

External power rail switching offers the best long-term leakage power savings – but introduces a significant turn-on delay to allow voltage regulation to stabilize and settle within specification.

Only a few voltage rails can typically be externally switched; every power supply incurs (external) regulator cost and area on the circuit board – including inductors and capacitors required to implement switched mode power supplies. Every power rail also requires on-chip power distribution that costs area and complicates the power planning and physical floor-planning. Most SOC's already have at least three power rails:

- IO power (at least one of 1.8/2.5/3.3V, and perhaps several depending on the application
- "Always-on" SOC core rail (technology dependent logic and internal memory power rail)
- Clean analog power supply rail to PLL's.
- An optional "keep-alive" voltage supply to the real-time clock

Adding more than two or three external switch power rails adds significant complexity and cost to the end-product.

Typically a shared ground/VSS connection approach to the chip and board works best for external power rail switching. Although there are typically independent VSS pins for both the IO pad-ring and the chip core to de-couple output simultaneous switching activity from the logic and memory, these are typically grounded on the circuit board into a shared "0-volt" ground plane. Treating any other power supplies as switched positive supplies relative to the common ground minimizes complexities when adding power gating.

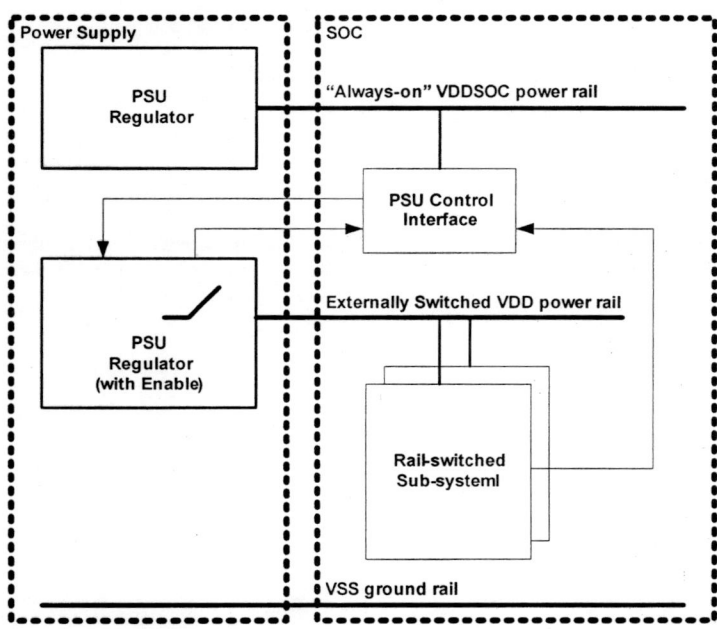

- Minimize the number of external switched independent power rails – each one must be justified from an end-product requirement given the associated additional power supply real-estate costs and on-chip power distribution.

- With external switched rails, it is best to switch (positive) supply rails and retain a common ground.

- In systems implementing voltage scaling an independent rail must be provided for each voltage scaled region.

Pitfalls:

- Design for significant external power rail switching times: tens or hundreds of thousands of clock cycle latencies must be factored into wake-up and will be dependent on the external PSU specifications.

- Although multiple rails appear elegant from a system design perspective they introduce verification and deployment challenges in production. Independent supply rails have independent voltage control regulators, and independent rails can exhibit vastly different load regulation characteristics when active, wait-stated or halted compared to logic powered at interfaces.

6.2.2   On-Chip Power Gating

On chip power gating is much faster than off-chip power rail gating. And the smaller the power gated region, the faster power can be gated on and off. The current required to power up a small power gated region is much less than that required for a large block. But time must be budgeted to manage the minimization of power gating transients and noise injection as seen by other logic and memory.

Therefore it is realistic to see power gating in terms of a few clock cycles for very small regions and tens or even hundreds of clock cycles for more significant gate counts. Turning on a number of small power-gated regions at the same time is no better than powering up a large block and may lead to a much more complex power controller.

Power gating has an impact on both performance and area, as will discussed later chapters, due to the nature of the switching transistor fabric. These limitations will impact system architecture and design objectives.

Recommendations:

External power rail switching incurs significant delays on wake-up events – from the order of tens of microseconds to milliseconds or even longer. Much faster supply switching times are not necessarily desirable. The in-rush currents to re-charge all the capacitive nodes in the powered-down subsystem result in noise injection into other (powered) regions of the chip. The resulting "ground-bounce" in a shared ground system can introduce problems that are hard to quantify until very late in the implementation and analysis phases of the design flow.

Translating such latencies into clock cycles at RTL level is not simple. Normally the clocks should be suppressed until a switched power rail is stable and within specified tolerance. For a design operating in the hundreds of MHz region this may be the equivalent of tens of thousands of clock cycles. The actual delays are highly dependent on the power supply technology (which may have to be multi-sourced in a production).

Separate power rails become a necessity when one introduces dynamic voltage scaling (Chapter 9). It may also be highly desirable to give large banks of memory their own supply which may be switched to intermediate RAM retention operating conditions, for example. This is discussed more in Chapter 13.

- Design for technology-dependent power-gating times: tens or hundreds of clock cycle latencies may need to be factored into wake-up times dependent on the area switched and the switching fabric control characteristics.
- Design for "wait-states" across boundaries where there are dynamically power gated functional units such that the implementation-dependent delay times can be safely managed and latency constraints set.

Pitfalls:

- Every power-gated rail introduces verification and test challenges so the number of power gated regions needs to be carefully justified and factored into project timescales.

6.3   Power State Tables and Always On Regions

When dealing with multiple power-gated power domains, power routing can become complex. In particular, the concept of "always on" becomes less clear. Figure 6-4 shows three power domains, each of which is power gated.

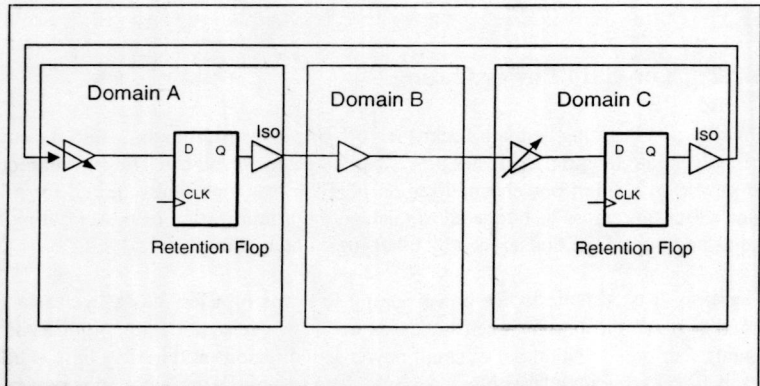

Figure 6-4   Buffering Inter-Domain Signals

If power domain B is always on, then there is no problem. But if domain B is turned off while domains A and C are powered up, then there is a problem: the outputs from A to C are corrupted because the buffer in B is powered down. In this case, we would have to route power from some other "always on" supply to the buffer in B. We could use either the isolation supply in A (since it stays on even when A is powered down) or the supply from C.

On the other hand, if we know that whenever B is powered down, then C is also powered down, we do not have to provide a special supply to B. In this case, we consider B to be "relatively always on" – that is, always on relative to domain C.

Thus, we can end up with some fairly complicated power routing rules depending on the power gating relationships among different blocks.

UPF provides a succinct way for system architects to communicate these power gating dependency rules to the implementation tools.

The **create_pst** and **add_pst_state** commands allow us to create a power state table that can be used to specify the relationships between different power supply nets. See Appendix B for a description of these commands.

CHAPTER 7

A Power Gating Example

The SALT technology demonstrator project provided a platform for testing the approaches to power gating and state retention described in this book. In this chapter we give some more details on the system design and RTL coding for this project.

The SALT chip is implemented in 90nm generic technology and contains an ARM processor, an AMBA bus and set of peripherals, and a Synopsys USB OTG digital core and PHY. The ARM core and the USB core are independently power gated. The ARM core uses full state retention; the USB uses partial state retention. Both the ARM core and the USB use switching fabrics of header switches; thus they switch VDD and use a common ground (VSS).

In this first section, we describe the power gating design for the processor. Figure 7-1 on page 86 shows a simplified block diagram of the SALT chip.

7.1 Leakage Modes Supported

Most battery-powered processor based designs have to deal carefully with the balance between performance (to support product features) and low power (to support long battery life). The performance requirement steered us to a higher performance, leakier process. To maintain long battery life, we needed to provide aggressive leakage management.

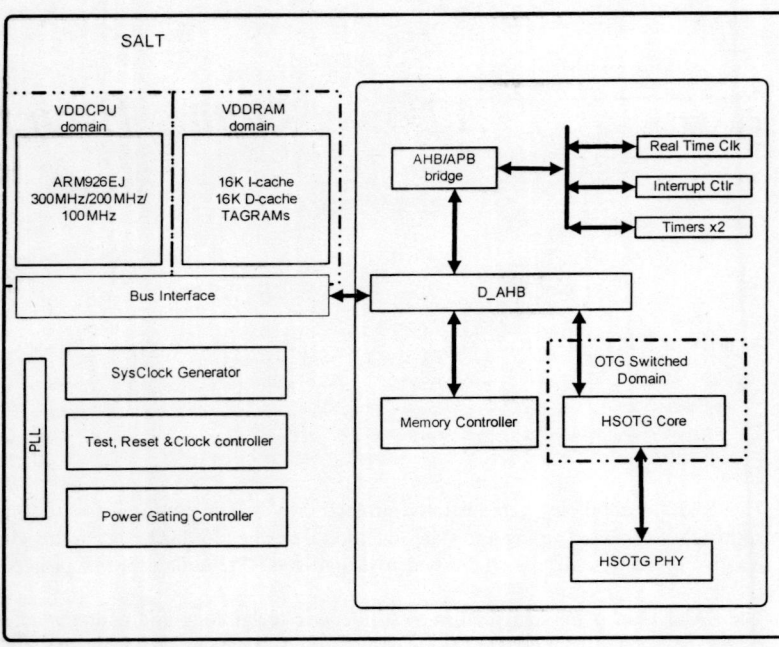

Figure 7-1   SALT Block Diagram

For the SALT project, the processor uses four low-power modes. In all modes, the power controller generates a **SLEEP** signal to enter the low power mode and the **WAKE** signal to exit. In order of increasing leakage savings – and increasing time to power up and power down – the modes are:

- **HALT**: SLEEP turns off the clocks to the processor, WAKE restarts the clocks. Power remains on.
- **SNOOZE**: SLEEP initiates power gating with state retention; power is switched off to the processor itself, but the cache memory remains powered up. WAKE initiates the power up and restore sequence; power is brought up as fast as safety allows.
- **HIBERNATE**: SLEEP initiates a sequence of scanning the processor's internal state to external memory; the VDDCPU power rail is then switched off. WAKE causes the power rail to be switched back on, and the scan chains are used to restore the internal state of the processor. A 32-bit AMBA-based bus is used to write to and read from an external memory. A 32-bit CRC provides integrity checking. The CRC is saved along with the scanned data and used to protect

against restarting with corrupted state. Note that VDDCPU provides power to the processor logic, not cache. So in HIBERNATE mode, only the processor is powered down; the cache memory remains powered up.

- **SHUTDOWN**: SLEEP initiates the same power down sequence as in Hibernate, but now both the processor and the cache have their external power supplies turned off. Explicit code must be called to write back any dirty data in the cache memories before both the VDDCPU and VDDRAM supplies can be power-rail switched. This mode is the only mode not transparent to the operating system.

In addition to these modes, the SALT chip provides active leakage reduction through externally managed threshold scaling. This threshold scaling is done using back-bias control. Both P- and N-wells for the CPU standard cell area are brought out to the chip's pins. This enables experimental analysis of delay and leakage power characteristics as the bias is varied.

Using this well-biasing scheme, we defined three operating modes:

- **NORMAL**: with standard well-bias
- **TURBO-RUN** – with the wells forward biased for increased speed
- **POWERSAVER-RUN** – with the wells reversed biased for reduced leakage

The power controller is designed to ensure that well-bias voltages are only changed while the design is static and un-clocked. Any change between "normal" and back- or forward biased modes of active power management go through one of the HALT, SNOOZE, or HIBERNATE states. Figure 7-2 on page 88 shows the state diagram for the power modes.

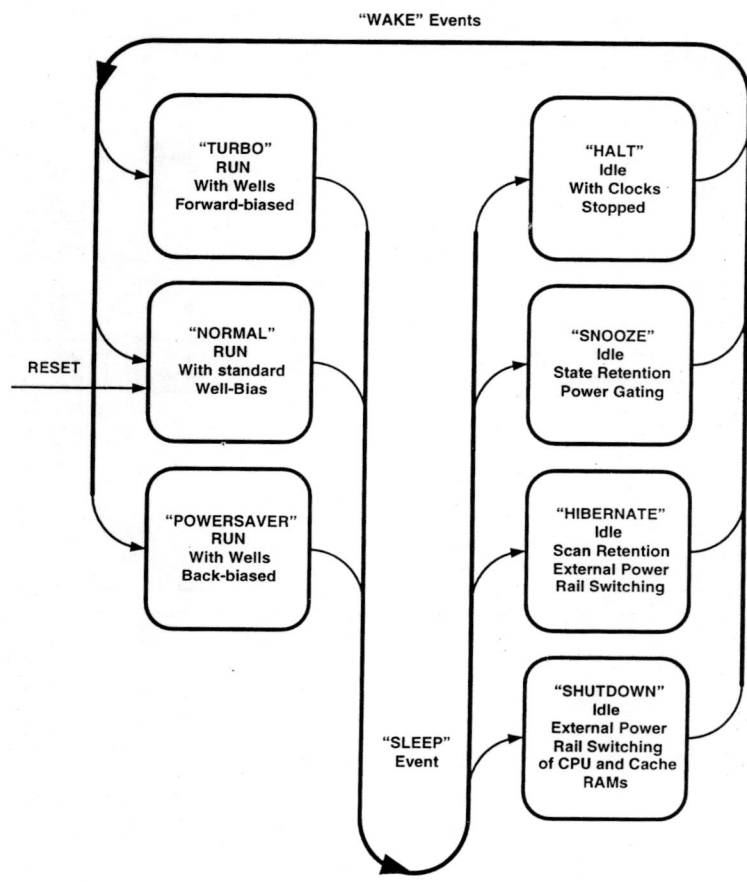

Figure 7-2   Power States for SALT

7.2   Design Partitioning

The RTL design is partitioned to allow the three primary power supplies to be mapped to the RTL design:

- **VDDSOC** is the "always-on" supply that powers the entire chip except the processor and its cache.

- **VDDRAM** is an external switched power rail that supplies the Cache and MMU RAMs.
- **VDDCPU** is an external switched power rail that supplies the CPU standard cell area.

VDDSOC provides power to the digital side of the PLLs, the clock generators and the power management control blocks, plus all the real-time peripherals. These peripherals include a real-time clock and timers; these can generate wake-up events as part of their interrupt service requests.

VDDSOC also provides power to the USB OTG subsystem. The USB is power gated though its own power switching fabric using a separate power controller.

Using a separate power supply (VDDRAM) for the cache accomplishes several objectives: it allows the CPU to be powered down while keeping the cache powered, allowing faster restore. It allows the CPU and cache to be powered down while keeping the peripherals of the chip powered up. These peripherals control the wake up of the CPU – upon detecting an interrupt, the interrupt controller signals the power controller to power up the CPU. Finally, using a separate cache power supply allows development of detailed leakage and active power consumption profiles, so that we can determine the minimum operating voltage for the cache memory and the minimum voltage that assures that the cache retains is contents during CPU shut-down.

Similarly, using a separate supply (VDDCPU) for the CPU logic allows detailed leakage and active power consumption profiles to be measured as well as the time and energy cost to get in and out of each power saving state. Figure 7-3 shows the partitioning of the SALT chip.

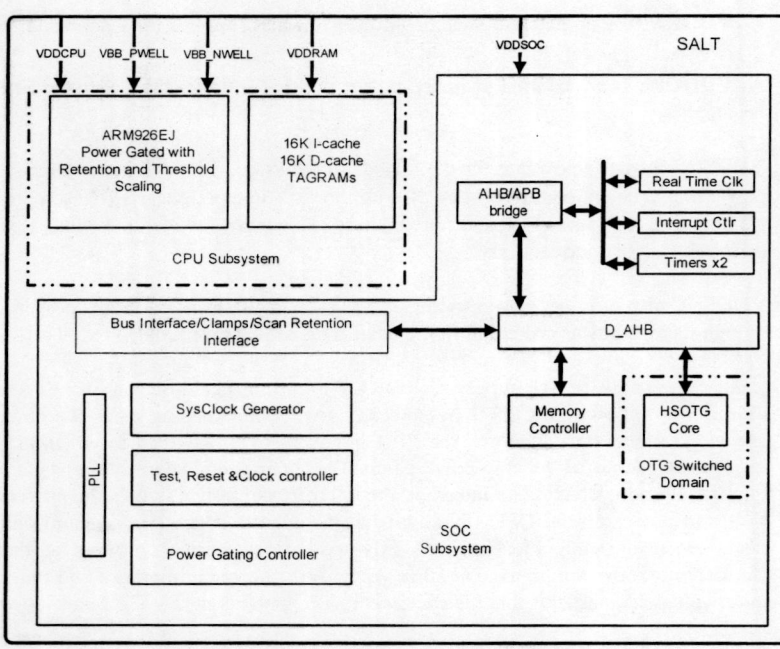

Figure 7-3   Partitioning of the SALT Chip

Since the original SALT design, the CPU sub-system has been re-implemented to improve design re-use. The Bus Interface Unit and the Power-Gating Controller are now integrated into a "VDDSOC" region within the CPU. This integration signifi-cantly reduces the number of signals – and blocks – that the SoC design team has to deal with when integrating the ARM core into the chip.

Although this makes the 3-supply-rail CPU subsystem slightly more complex to implement, the timing and internal power gating and isolation interfaces are now all abstracted away from the top-level SOC design. Any changes or enhancements to the low-power states supported by the IP block are independent of the top level system design.

Figure 7-4 shows the re-partitioned CPU subsystem.

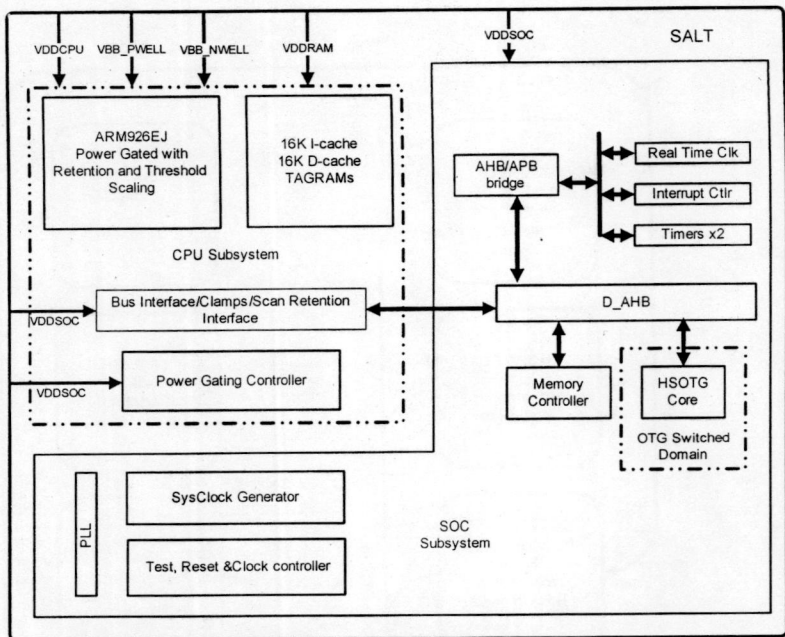

Figure 7-4   Re-Partitioned CPU Subsystem

Power Gating control and handshakes

The CPU power-gating control system manages:

- the interfaces to the external power supplies for hibernate and shutdown modes
- the bus-clock synchronous scan clock pulses for hibernation-mode save and restore.
- local header-switch power gating control for snooze mode
- isolation and state retention for snooze mode
- the handshake with the system clock generator to switch CPU clock frequencies for the different run modes

One of the goals of the SALT project is to develop a methodology that supports a wide range of library components and current in-rush management techniques. To do this, the CPU power controller uses a request-acknowledge handshake for every con-trol signal: power gating, isolation, save, restore and reset.

All the acknowledge signals are treated as asynchronous to the controller clock, and have local synchronizers to the state machine clock domain.

This approach ensures that the design is free of locally coded delays or counts. It also allows the acknowledge signals to be tied directly to the requests for some implementations or built as true handshakes. In the current SALT implementation, the acknowledges for isolation, save and restore, and reset, are all tied to their respective requests; as implemented, these functions are all very fast and require no delay in sequencing.

The power gating control of the header switches uses a true acknowledge. The power down request is daisy chained through the header switches, so that the acknowledge that arrives back at the controller has a delay equivalent to the switching time of the whole fabric.

A couple of additional design notes:

- After asserting reset, the initialization (power on reset) sequence pulses the save/restore signaling to flush out any X-s from the shadow retention flops. This may be useful when running functional test programs or vectors on the gate level netlist.
- All timing-dependent state machine transitions include a holding term that waits for the output asserted in that state to be acknowledged in order to maintain the timing-independent request/acknowledge sequencing
- The power-gating assertion and de-assertion use a request-acknowledge handshake.
- In the SALT project there are extra diagnostic control inputs to control the switch fabric which allow the power gating to be soft-sequenced or forced fully on and off, and only the power-gating acknowledge input to the state machine is used to determine when power is safely restored.

7.3   Isolation

The SALT project uses several different isolation techniques. The initial version of SALT was done before there was tool support for automatically inserting isolation cells. As a result, we manually inserted these cells in the RTL.

The interface between the CPU and cache is particularly timing critical, requiring careful design and timing analysis. We put isolation cells on all of the outputs from the CPU to the cache, but placed them in the cache's VDDRAM region. This was convenient in the SALT design because the cache is always powered up, making power routing to the isolation cells simpler.

The VDDRAM region has input isolation cells instantiated as "Generic Library Cells." These are just RTL wrappers that can be mapped either to behavioral simulation models or to technology-specific clamp cells from the "Power Management Kit" for the standard cell library. This approach provides explicit instantiation of the cells on the many critical path signals, including clocks and resets, into the memories from the CPU core logic. Instantiating these cells in the RTL provides clean visibility of the clocks and critical signals that need to be managed carefully in the implementation flow. Clock balancing across isolation cells is not straightforward; isolation cells typically limit the flexibility the clock buffering tools in restructuring the buffer trees.

The outputs from the CPU to the Bus Interface Unit are isolated by cells placed in the CPU – that is, in the VDDCPU region. These output isolation cells pull-down all output signals at the AMBA bus interface in Snooze mode (local power gating), to guarantee clean SOC interface signals.

However when the CPU rail is switched off (Hibernate mode) these isolation cells lose their VDDCPU power and the outputs could again float. To address this problem, simple bus repeater or hold cells were added in the VDDSOC domain to clamp these outputs while avoiding any further gate delays. In addition, the isolate control signal from the power controller drives reset in the bus interface module to force logic-0 clamping of all bus interface protocol signals.

The power controller generates a single isolate control signal, but it is routed as two separate signals. On copy goes to the cache RAM without going through the CPU region, and one copy goes to the CPU without going through the cache. When VDDCPU has its power rail shut down during hibernate, the isolate signal for the cache is still powered up since all the buffers in its path are powered by VDDSOC or VDDRAM. Thus, the inputs of the cache are protected from floating signals (and data corruption) during hibernate.

The USB OTG block uses the alternative of instantiated AND gate cells in the RTL with suitable "don't-touch" attributes added to prevent logical optimization across these isolation boundaries. During synthesis we force these AND gates to be mapped to cells that do not crowbar with one floating input (as long as the other input is low).

All of the isolation cells in the SALT design were instantiated using a generic, technology-independent wrapper module. For hand instantiated cells, this provides a degree of design portability and enables simulation before a specific technology library has been selected.

The initial version of the SALT chip instantiates isolation cells in the RTL as described above. In a subsequent version, we use EDA tools to implement the isolation cells without modifying the RTL. The re-partitioning of the CPU subsystem to include VDDSOC also eliminates the need for bus repeater cells for the AMBA bus interface signals. Now the isolation cells get power from VDDSOC, which is always on. Thus, a single set of isolation cells is used to isolate the bus interface signals.

With the UPF-defined simulation semantics, it is possible to instantiate and simulate isolation cells by issuing a UPF tcl command (set_isolation); no hand instantiation or generic wrappers are required.

7.4   Retention

The SALT project incorporated several retention techniques to allow comparison and analysis of the area/time/energy cost functions for each technique on the same silicon.

The CPU uses full state retention. Given a fully validated CPU core, retaining every register bit state is the only safe approach to ensuring that the processor can be restarted with arbitrary control and data state. To retain only the architectural state, and use reset signal for non-architectural state, would require a serious verification project.

A key concern of any power gating design is to assure that retention registers do not get corrupted during the power down/power up sequence. On SALT we added a non-real-time diagnostic mechanism to the power control sequencing. It uses the "Hibernate" scan functionality to check-sum and save to external memory the contents of the all the shadow registers after the SAVE operation and then checksum and save the entire register contents after the RESTORE operation. These can then be compared to detect any corruption of the retention registers during power sequencing.

This approach allows error analysis for both random and location sensitive problems and the efficacy of the soft-start power gating sequencing in limiting in-rush current. This also turned out to be a valuable way of quantifying the safety margins of the retention flops and allowed them to be subjected to thermal and voltage shocks while in retention mode.

On the other hand the USB OTG core uses partial retention. The core is partitioned with persistent USB end-point data held in the Control and Status Register (CSR) block, while the Protocol Interface Engine (PIE) block contains only protocol state. Data for the current transaction is held in a RAM-based FIFO. During power down, the USB waits for the current transaction to finish, so the FIFO is empty. It then saves all the state of the CSR, using the standard retention register technique. On power up, the CSR has its state restored, and a reset is issued to the PIE. This resets the protocol engine, and it is ready for the next transaction.

The USB power gating design is discussed in more detail in Chapter 8.

7.5   Inferring Power Gating and Retention

In the SALT project, we use a retention register cell with a single-pin control to control save and restore in an edge-triggered manner. Retention state is captured on the falling edge of an active-low **NRETAIN** signal and restored on rising-edge of the same **NRETAIN**:

Building on the same worked example described in Chapter 5 we can modify the UPF to show this retention behavior as follows:

```
set_retention_control my_retention_strategy
    -domain my_power_domain
    -save_signal {NRETAIN negedge}
    -restore_signal  {NRTAIN posedge}
```

Or, if we are using a simulator that does not support UPF, we can modify the RTL to show retention behavior as follows:

```
`ifdef  RTL_PG_EMULATE
    reg [3:0] state_SAVE = 4'bXXXX;
    wire PWR;
    assign PWR = pwr_req & pwr_ack;
`endif

always @ (posedge clk or negedge nrst
`ifdef  RTL_PG_EMULATE
    or negedge PWR or negedge NRETAIN
`endif

) begin
`ifdef  RTL_PG_EMULATE
    if (!PWR)
        current_state  <= 4'bXXXX;
    else if (!NRETAIN)
        current_state <= state_SAVE;
    else
`endif

    if (!nrst)
        current_state  <= 4'b0101;
    else
        current_state <= next_state;
end
```

```
`ifdef  RTL_PG_EMULATE
always @ (negedge NRETAIN)
    state_SAVE <= current_state;
`endif
```

7.6   Measurements and Analysis

We evaluated the SALT project silicon to understand the power gating and state retention improvements to leakage power, and the effect of the switching fabric on the functional performance.

Figure 7-5 shows the basic evaluation of the dynamic and leakage power as measured for both the cache memories and the standard cell logic for the ARM926EJ CPU subsystem stepping the power supply in 10% steps from 110% of nominal down to 70% of nominal. For this 90nm Generic process technology the nominal supply voltage is 1.0V.

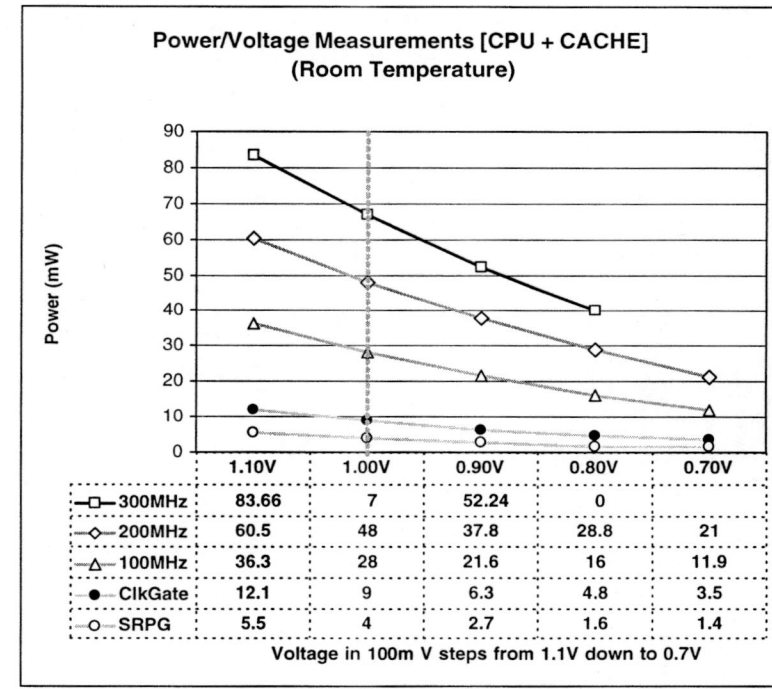

Power/Voltage Measurements [CPU + CACHE]
(Room Temperature)

	1.10V	1.00V	0.90V	0.80V	0.70V
—□— 300MHz	83.66	7	52.24	0	
—◇— 200MHz	60.5	48	37.8	28.8	21
—△— 100MHz	36.3	28	21.6	16	11.9
● ClkGate	12.1	9	6.3	4.8	3.5
—○— SRPG	5.5	4	2.7	1.6	1.4

Voltage in 100m V steps from 1.1V down to 0.7V

Figure 7-5   Power Measurements for SALT

The measurements are at the circuit board level – before any package/bond-wire/on-chip IR drop and the power gating fabric itself – so are very reasonable for "typical" silicon at room temperature.

The worst-case timing sign-off target frequency for the CPU design was 300MHz. The first three measurements show the dynamic power with the CPU operating at 300MHz, 200MHz, and 100MHz respectively.

The "ClkGate" measurements reflect the baseline leakage when the clock for the entire CPU is gated off.

The "SRPG" (Save Restore Power Gating) measurements are the leakage during power gating. Thus it reflects the leakage of the switches, the retention latches and the always-on control buffer trees, and in this case also includes the baseline leakage for the cache memories that are not power-gated.

Figure 7-6 shows the measured leakage power for the power-gated "VDDCPU" power domain across the temperature range of key interest for battery-powered products. The vertical scale is plotted logarithmically:

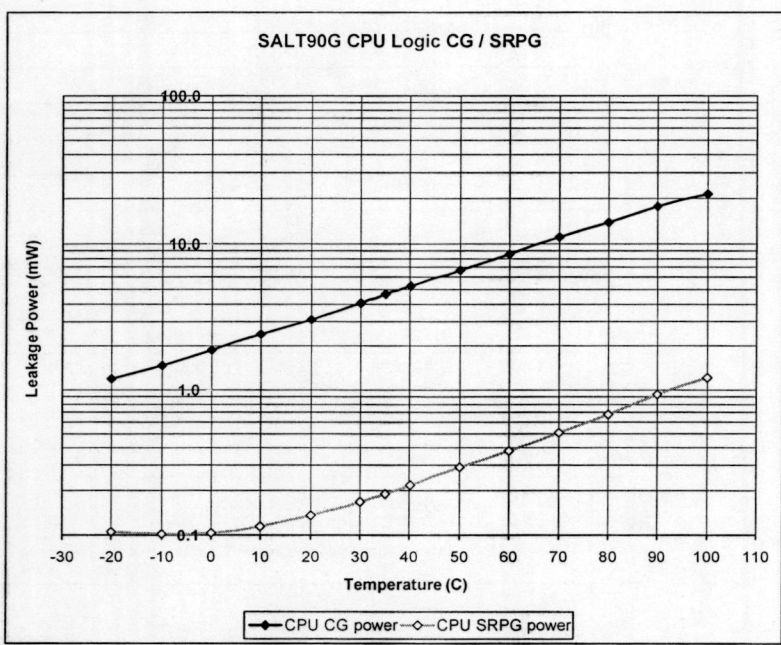

Figure 7-6   Leakage Power on SALT

The upper curve plotted is the baseline leakage when the clocks are stopped. The lower curve is the measured leakage power with State-Retention Power-Gating. Across the entire temperature range the leakage power savings are more than 10-fold and greater than 25-fold around room temperature.

Figure 7-7 shows the measured leakage power for not only the power gated standard cell logic but also the cache memories, which were not available with power-gating for this project:

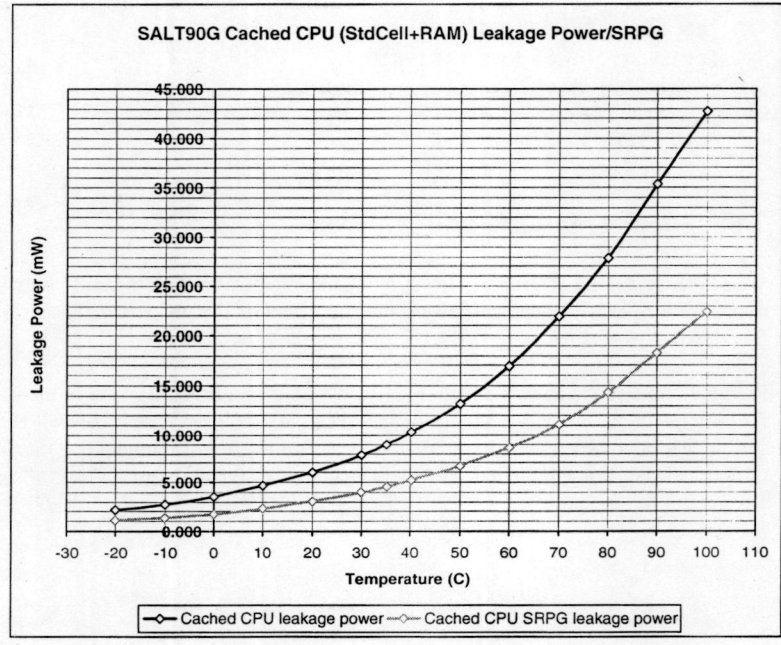

Figure 7-7   Leakage vs. Temperature

The upper curve is the baseline total leakage for the CPU subsystem (logic plus RAMs) and the lower curve shows the power with the logic portion power gated with state retention.

The leakage power of the RAM now dominates the leakage power but there is still a factor of two leakage power saving for the CPU subsystem. With integrated power gated RAMs the savings could be improved to closer to those achieved for the logic portion.

Figure 7-8 on page 100 shows the detailed comparison of the leakage power savings of state retention relative to the baseline leakage. The ratio of baseline leakage to SRPG leakage is plotted, on a linear scale, to show how this characteristic varies across the temperature range.

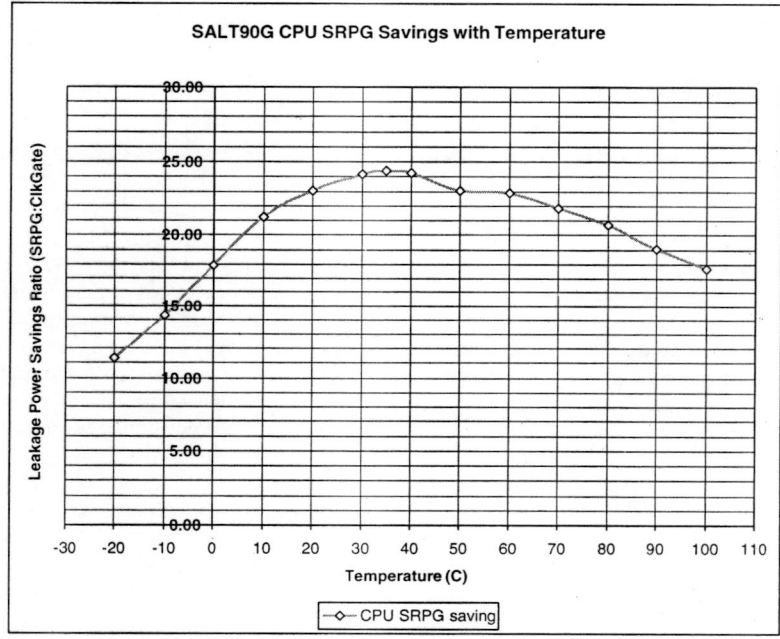

Figure 7-8　Leakage Savings vs. Temperature

The deterioration of the savings at the high end of the temperature range can be explained by the leakage behavior of the High-V_T switches. The very best power-gating leakage power saving for this particular design and technology was achieved at 35C.

In summary, power gating techniques when applied to standard-cell logic produced leakage power savings of between 10x and 25x over the baseline leakage. Because cache memories are typically tuned for performance and so exhibit a fairly high leakage power, these too benefit from leakage reduction techniques. This topic is addressed in Chapter 12.

The SALT technology demonstrator has been an effective vehicle for developing State-Retention Power-Gating techniques and methodologies and for analyzing the costs and benefits of power gating on a 90nm process. These techniques and methodologies will be highly appropriate at 65nm and below.

CHAPTER 8　　　　　　　　*IP Design for Low Power*

The previous chapters have discussed low power design from the perspective of the system architect and chip designer. This chapter describes low power design from the perspective of the engineers who design complex IP, such as processors, DSPs, USB, PCI Express, and bus infrastructure. Until now, we have assumed that the IP is relatively fixed, and that we must add low power capability to it. Now we discuss how to design complex IP to meet our low power objectives.

Today the vast majority of complex chips are designed using IP – either third party or internally developed. And the key to designing good IP is to design it in a way that allows it to be used in multiple applications.

To assure that an IP can be used effectively in multiple applications requiring low power, we must design it so that it can be used with different power strategies. In one application, clock gating and multiple V_T libraries may provide low enough power. In other applications, aggressive on-chip power gating may be required. In other applications, dynamic voltage scaling may be the key to achieving the chip's power goals.

To address these various needs, we need to do the following:

- Partition the design to support various low power strategies, especially power gating
- Include explicit support for power gating
- Develop reference power intent files
- Design the clocking and reset strategy with low power in mind
- Package the IP to support low power
- Verify the IP using various low power strategies

As we go through the architecture, design, and packaging process for IP, we need to bear in mind that any or all of the following techniques may be used in the actual implementation of the IP in a chip:

- Multi V_T
- Clock gating
- Power gating (internal and/or external)
- Voltage Scaling

For certain types of IP, different kinds of support for these functions may be required.

Memory and other hard IP blocks have special requirements for low power. Low power memories often have various modes: a normal operating mode, a retention mode, and power off. In retention mode, the voltage is lowered to the minimum required to retain data, but below that required to do reads and writes.

Physical layer interfaces for IO standards like USB or PCI Express typically have more than one power mode as well. In addition to the normal operating mode, there may be a complete shutdown mode that drives power close to zero. There may be an additional operating mode where enough of the circuit is powered up so that it can wake up in response to activity on its interfaces.

Configurable soft IP, because it can be configured by the user, offers a complex design challenge that in some sense is a super set of the challenges of hard IP. For soft IP, multiple power modes and multiple power reduction techniques must be supported in a user-configurable way that is robust, easy-to-use, and flexible. How to design such IP is the focus of the rest of this chapter.

We use as an example the USB On-the-Go (OTG) IP that we included on the SALT chip. The OTG core is a Synopsys digital soft IP core that was designed with power gating in mind, but which had never actually had power gating capability added to it. We modified the RTL to add a power gating controller, retention registers, and isolation cells. Today, of course, we would use UPF to describe most of these modifications, as described later in this chapter.

8.1   Architecture and Partitioning for Power Gating

A good working definition of architecture (in the context of IP design, at least) is the partitioning and interface design of the IP. In supporting various low power strategies, power gating presents the most significant new architectural challenge in the architecture of IP.

To support power gating, we need to:

- Decide when and how the IP will be powered down and powered up
- Decide which blocks will be power gated and which blocks will be always on
- Design a power controller that controls the power up and power down sequence
- Determine which signals need to be isolated during power down
- Develop an initial strategy for clocks, resets, and the power control signals

8.1.1   How and When to Shut Down

On the SALT chip, we included a power gated versions of the CPU and the USB OTG digital core. The strategy for the CPU was to have the power down sequence controlled by software. When the software determines that it wants to power down the CPU, it signals the CPU power controller. The controller then goes through the power down sequence. Just enough of the processor is kept alive to respond to an interrupt. When the appropriate interrupt occurs, say from a timer peripheral or from an external source, the power controller for CPU then goes through the power up sequence.

The strategy for the USB OTG was to power down during idle times, but only when allowed by the CPU. The CPU writes an enable bit in a register in the USB OTG to enable power down – essentially saying that it is done with transactions. The USB OTG then waits until the USB bus has been idle for 3ms (indicating that the USB OTG can enter SUSPEND mode). On entering SUSPEND, the USB signals to the USB power controller to begin the power down sequence. Enough of the USB OTG is kept alive that it can respond either to a read/write from the CPU or to activity on the USB bus. If the CPU clears the power down enable bit, or if there is activity on the USB bus, then the controller goes through the power up sequence.

8.1.2   What to Shut Down and What to Keep Alive

Figure 8-1 on page 105 shows a simplified diagram of the USB OTG digital core. During power down, the Bus Interface Unit is kept powered up so that it can respond to a CPU request to power up. Similarly, the PHY Interface block is kept powered up so that if USB activity is detected, it can signal the power controller and wake up the core. The Clocks and Reset block is also kept powered up to provide clocks to the Bus Interface Unit and the PHY Interface.

All of the rest of the USB OTG core is power gated. The status and control registers are saved and restored in the power gating sequence, using retention registers controlled by a single pin, NRETAIN. The protocol engine, since it starts each transaction from scratch, is simply reset at power up.

The power controller is included in the AHB Slave block of the Bus Interface Unit simply as a matter of convenience – it runs off the AHB clock, needs to stay powered up during power gating, and the power gating enable register is located there as well. But the controller could just as easily have been a separate block.

There are two clock domains in the power gated region of the USB OTG digital core – the AHB clock domain and the PHY clock domain. Synchronizers are used for control signals, including power gating control signals, that pass between the two domains. Thus, the time from the assertion of a power gating control signal (in the AHB domain) to the time it affects the PHY domain is not deterministic. And in fact it can vary greatly, as the timing relationships between these two domains can vary greatly from application to application.

For this reason, request-acknowledge handshakes are used for a number of the power gating control signals.

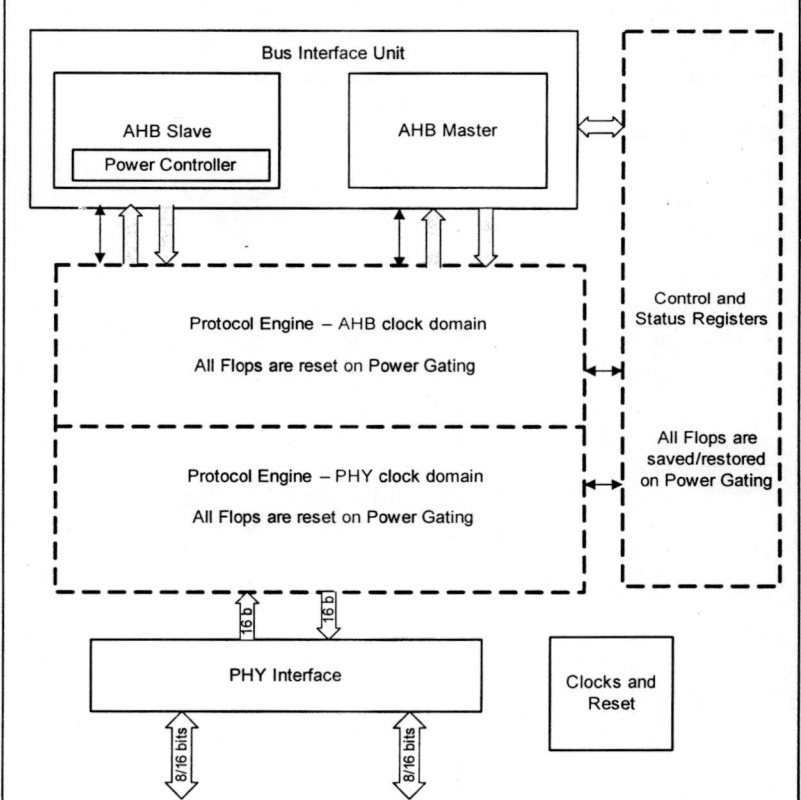

Figure 8-1   USB OTG Block Diagram

8.2   Power Controller Design for the USB OTG

The power controller in the USB OTG is conditionally compiled using `ifdef; if the user does not want to use power gating, the controller is not compiled into the design.

The power controller is a simple state machine that controls the following signals:

- pwr_reset_n   // the reset to the protocol engine
- gate_hclk,   // control signal to turn off clocks in the AHB domain
- h2pd_stop_pclk   // control signal to turn off clocks in the PHY domain

- bius_pwr_clamp   // control signal to clamp outputs of the AHB domain
- h2pl_pwr_clamp   // control signal to clamp outputs of the PHY domain
- pwr_dwn_req_n   // control signal to request power down (active low)
- retain_n   // negative edge is save; positive edge is restore

And it receives the following inputs:

- pwr_dwn_ack_n   // acknowledge for pwr_dwn_req
- stop_pclk_ack   // acknowledge for h2pd_stop_pclk
- pwr_clamp_ack   // acknowledge for h2pl_pwr_clamp
- suspend_detected   // indicates no USB activity for 3ms
- fifo_flushed   // indicates all pending USB transactions are done
- wkup_res_det   // indicates activity has been detected on the USB bus
- enable_power_gating   // from CPU to enable power gating

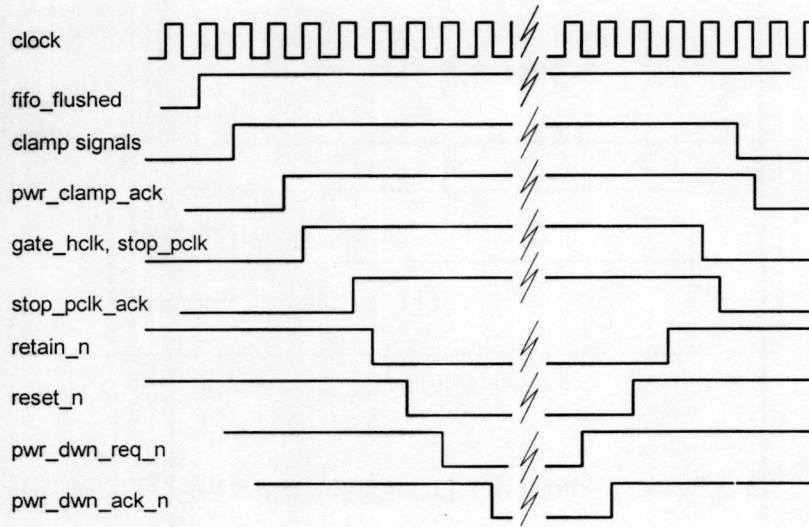

Figure 8-2   Power Gating Sequence for the USB OTG

When the power controller sees that suspend_detected is asserted (and the power gate enable bit is set in the status register), it starts the power down sequence. This sequence is shown in Figure 8-2 and described below:

- It waits until fifo_flushed is asserted. This indicates that all pending transactions have completed, and the fifo that stores pending transactions is empty.
- It then asserts bius_pwr_clamp (to the AHB clock domain) and h2pl_pwr_clamp (to the PHY clock domain) to clamp the outputs of the power gated portion of the USB OTG.
- It then waits for pwr_clamp_ack (from the PHY clock domain). This tells the controller that the isolation cells are all clamped (in the USB OTG, they are all clamped to "0.")
- It then asserts gate_hclk and h2pd_stop_pclk to stop both AHB and PHY clocks.
- It then waits for stop_pclk_ack (from the PHY clock domain) to indicate that clocks have stopped in the PHY domain. Because of synchronizers, the PHY clocks always shut down after the AHB clock.
- It then asserts retain_n (an asynchronous signal, so no handshake required). This causes the retention registers to save their contents.
- It then asserts reset_n (an asynchronous signal, so no handshake required).
- It then asserts pwr_dwn_req_n, causing the power gated sections of the USB OTG to be powered down.
- It then waits for pwr_dwn_ack_n to be asserted, indicating that the USB OTG is completely powered down. Once this acknowledge is received, the power controller goes to an idle state, waiting to be told to wake up the USB OTG.

When the power controller sees either that enable_power_gating has been cleared (indicating that the CPU wants to power up the USB OTG) or that activity has been detected on the USB bus, then the power controller state machine starts the wakeup sequence:

- It de-asserts pwr_dwn_req_n, causing the power gated sections of the USB OTG to be powered up.
- It then waits for pwr_dwn_ack_n to indicate that the USB OTG is completely powered up.
- It then de-asserts reset_n, so all the flops in the protocol engine resume in the reset state.
- It then de-asserts retain_n so all the retention flops in Control and Status Register block are restored.
- It then de-asserts gate_hclk and h2pd_stop_pclk to start both clocks.
- It then waits for stop_pclk_ack to indicate that the clocks are running.
- It then de-asserts bius_pwr_clamp and h2pl_pwr_clamp to release the clamps on the outputs of the power gated portion of the USB OTG.
- It then waits for pwr_clamp_ack. This tells the controller that the isolation cells are all released.

Once this acknowledge is received, the power controller goes to an idle state, and the USB OTG resumes normal operation.

8.3   Issues in Designing Portable Power Controllers

In addition to implementing the functions described above, the power controller for an IP must be designed to be portable - that is, to be used in multiple applications and with different libraries. The two major challenges here are:

- Dealing with different libraries that may require different control signals
- Accommodating the (potential) need to interface with a system level power controller

The basic control functions are to turn power on and off, turn clocks and reset(s) on and off, turn isolation cells on and off, and issue save/restore commands to retention flops. These are common to virtually all power gating designs. But specific libraries may require different specifics:

- Signal polarity may be different for different libraries.
- Request/Acknowledge may or may not be required on any of the individual controls.
- Save and restore can be implemented either as a single control (retain_n in the example above) or as two separate signals.

Recommendations:

- Parameterize signal polarity on all control signals, so it can be configured by the user
- Implement a request/acknowledge handshake on all controls, but provide a mechanism so that the user can configure the IP to connect acknowledge to request for those applications that do not require a handshake.
- Parameterize the save and control function so that the user can configure it to be either a single control or dual (save and restore) control.

As power gating becomes more common, it is likely that chips will use a central power controller to coordinate the activities of the various power gated blocks in the design. In particular, if multiple blocks want to power up at the same time, it may be necessary to sequence through the blocks, powering up one at a time, to limit noise from excessive voltage spikes. A central agent may be needed to arbitrate among blocks wanting to power up or power down.

The kind of power controller described in the previous section should be able to accommodate such a system architecture. The user can route the power down request

to the central controller which can issue an acknowledge when it decides to service the request. The only requirement on the IP design is that the power down request and acknowledge signal must be available at the top level port of the IP.

8.4   Clocks and Resets

Clocks and resets for an IP need to be controlled for various reasons:

- For scan, normal clocks may have to be muxed with a scan clock
- For scan, resets may need to be controlled from a scan control pin at the chip level
- For power gating, clocks may need to be turned off and on for the power gated region while staying always on for the non-power gated region
- For power gating, resets may need to be selectively asserted to portions of the power gated region while not being asserted in other regions of the design

For these reasons, it is increasingly important to have a dedicated clock and reset module in the IP that handles only clocks and resets, and which provides the flexibility to meet the needs outlined above.

To provide optimal controllability for scan testing, we recommend that the power controller itself be controllable from outside the IP. That is, a central, chip-level scan controller needs to be able to force power on or off for the IP, as well as force clocks, reset, save and restore for the powered down region.

In the SALT chip, we also made the decision to do all scan clock muxing external to the USB OTG IP. The PHY clock is generated by the USB PHY and goes through a mux before providing the PHY clock to the digital core. This mux switches between the PHY clock (used by the digital core during normal mode) and the scan clock (used by the digital core during scan testing).

8.5   Verification

Verification for any configurable IP is a major challenge. This challenge is increased by adding power gating.

In developing the power-gated version of the USB OTG, we initially did full RTL functional testing without the power gating circuitry in place. Once the USB OTG passed all diagnostics, we added the power gating functionality and re-ran the diagnostics. Once these all passed, we ran a set of diagnostics for testing the power down feature itself.

Since the power gating function is completely independent of other USB functions, this approach seemed the appropriate way to minimize the verification effort and still provide strong verification.

At the RTL level, we simulated the power switching fabric by forcing register outputs to "X" during power down. This approach allowed us to completely verify the core at the RTL level. But because power gating is so closely tied to the physical implementation of a switching fabric, we also did extensive gate-level simulation of the power down function. This allowed us to use a detailed model for the switching fabric, including the time it takes to power up or down completely.

8.6   Packaging IP for Reuse with Power Intent

The SALT chip was developed before UPF was available to provide a convenient way of specifying the power strategy for an IP block. With the introduction of UPF, it becomes much more straightforward to include power strategies in the final packaging of the IP.

Any IP needs to be packaged in a way that enables users to configure the IP to their application. Often this is done using a configuration tool. For soft IP, the final packaging includes:

- Configuring the RTL (for the USB, this includes selecting the number of endpoints and configuring each endpoint)
- Ability to generate a test bench for the verifying the configured core, both pre and post-synthesis
- Synthesis scripts for the configured core, including support for clock gating and multi-V_T

For cores that support power gating, we need to add:

- The ability to configure the power controller
- The ability to generate a test bench for verifying pre and post-synthesis power operation
- The ability to configure the power intent, including the target retention registers
- Synthesis scripts that support the power intent
- Configurable UPF code to support the configurable power strategies

8.7   UPF for the USB OTG Core

Figure 8-3 on page 112 shows a more detailed block diagram of the USB OTG digital core. In the original SALT chip, we had to add retention, isolation, and the power switch directly in the RTL. With the introduction of UPF, we can describe this functionality using UPF tcl commands.

Note that the Power Control block still has to be designed in RTL and instantiated in the design.

The UPF code for adding power gating, retention, and isolation is shown below. The block names otg (top level of the core), biu, mac, etc., are the names of the instances of these modules in the RTL.

To make the UPF code portable, we use a variable ($otg) to indicate the actual path from the top of the design to the top level of the core in the RTL. We also use variables for the origin of the VDD and VSS power nets. Most likely these would be at the top level of the chip.

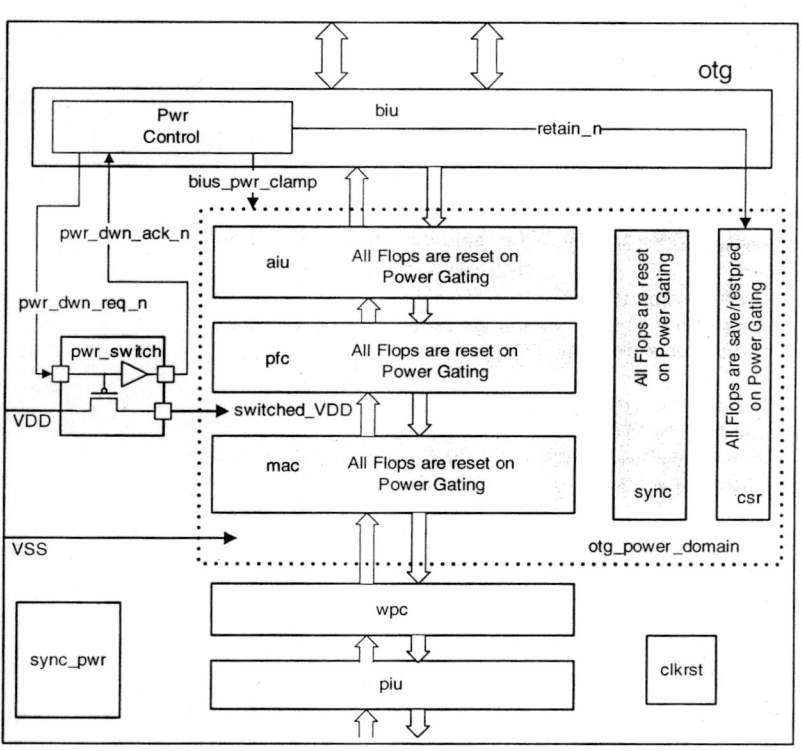

Figure 8-3   Power Gating for the USB OTG

```
set_scope $otg
create_power_domain otg_power_domain
    -elements {aiu pfc mac sync csr}

create_supply_net switched_VDD
    -domain otg_power_domain

set_domain_supply_net otg_power_domain
    -primary_power_net switched_VDD
    -primary_ground_net /$top_VSS
```

```
create_power_switch power_switch
    -domain otg_power_domain
    -input_supply_port {sw_input_port  /$top_VDD}
    -output_supply_port {sw_output_port  switched_VDD}
    -control_port {sw_control_port  biu/pwr_dwn_req_n}
    -ack_port {pwr_ack_port biu/pwr_dwn_ack_n}
    -on_state {pwr_on_state sw_input_port
               {sw_control_port ==1}}
    -off_state {pwr_off_state {sw_control_port ==0}}

set_isolation otg_isolation -domain otg_power_domain
    -isolation_power_net $top_VDD
    -clamp_value 0

set_isolation_control otg_isolation
    -domain otg_power_domain
    -isolation_signal biu/bius_pwr_clamp

set_retention otg_retention
    -domain otg_power_domain
    -retention_power_net $top_VDD

set_retention_control otg_retention
   -domain otg_power_domain
   -save_signal {biu/retain_n negedge}
   -restore_signal {biu/retain_n posedge}
```

8.8   USB OTG Power Gating Controller State Machine

The following code is included as an example of a simple power controller state machine. It is coded as a hierarchical state machine – a format which we find quite useful, particularly in more complex state machines.

```
//============================================================
//
//
//              (C) Copyright 2004-2005, Synopsys, Inc.
//                    ALL RIGHTS RESERVED
//
//
// Filename     : optc_sm.v
// Author       : Mike Keating
// Date         : November 28, 2005
// Version      : 1.0
// Description  : This Module implements state machine
//                part of power down control logic for
//                OTG
//
//============================================================
//============================================================

reg [2:0] main_state;
parameter TOP_IDLE= 3'd0;
parameter SLEEP=     3'd1;
parameter WAKEUP=    3'd2;
parameter FLUSH_FIFO=3'd3;
reg [2:0] SLEEP_state;
parameter SLEEP_IDLE=3'd0;
parameter CLAMP=     3'd1;
parameter SAVE =     3'd2;
parameter PWR_DOWN= 3'd3;
parameter CLOCKS_OFF=3'd4;
parameter SLEEP_DONE=3'd5;
parameter RESET_PDN =3'd6;
reg [2:0] WAKEUP_state;
parameter WAKEUP_IDLE=3'd0;
parameter WAKEUP_DONE=3'd1;
parameter PWR_UP=    3'd2;
parameter CLAMPS_OFF=3'd3;
```

```
parameter RESTORE=   3'd4;
parameter CLOCKS_ON=3'd5;
parameter RESET_OFF=3'd6;

always @ ( posedge hclk or negedge hreset_n )   begin
    if (!hreset_n)   begin
        bius_pwr_reset_n <= 1'b1;
        pwr_clamp_n_tmp <= 1'b1;
        bius_pwr_clamp_n_tmp <= 1'b1;
        h2pl_pwr_clamp_n_tmp <= 1'b1;
        bius_pwr_clamp_tmp <= 1'b0;
        bius_gate_hclk_tmp <= 1'b0;
        h2pd_stop_pclk <= 1'b0;
        retain_n <= 1'b1;
        pwr_dwn_req_n <= 1'b1;
        main_state <= TOP_IDLE;
        SLEEP_state <= SLEEP_IDLE;
        WAKEUP_state <= WAKEUP_IDLE;
    end else begin
        case (main_state)
            TOP_IDLE: begin
                if (suspend_detected_interrupt &&
                              enable_power_gating)
                  begin
                    main_state <= FLUSH_FIFO;
                  end
            end
            SLEEP: begin
              sleep;
              if (SLEEP_state == SLEEP_DONE &&
                  (sp2ht_wkup_res_det_biu ||
                   !enable_power_gating ))
                  begin
                    main_state <= WAKEUP;
                    SLEEP_state <= SLEEP_IDLE;
                  end
            end
            WAKEUP: begin
              wakeup;
              if (WAKEUP_state == WAKEUP_DONE)
                  begin
```

```
                    main_state <= TOP_IDLE;
                    WAKEUP_state <= WAKEUP_IDLE;
                  end
              end
              FLUSH_FIFO: begin
               . if (fifo_flushed)
                    begin
                      main_state <= SLEEP;
                    end
              end
          endcase
        end
    end

//----------------------------------------------------
--
//                      sleep_task
//----------------------------------------------------
task sleep;
  case (SLEEP_state)
      SLEEP_IDLE: SLEEP_state <= CLAMP;
      CLAMP: begin
        pwr_clamp_n_tmp <= 1'b0;
        bius_pwr_clamp_n_tmp <= 1'b0;
        h2pl_pwr_clamp_n_tmp <= 1'b0;
        bius_pwr_clamp_tmp <= 1'b1;
        if (pwr_clamp_ack_sync==1)
            SLEEP_state <= CLOCKS_OFF;
      end
      CLOCKS_OFF: begin
        bius_gate_hclk_tmp <= 1'b1;
        h2pd_stop_pclk <= 1'b1;
        if (stop_pclk_ack_sync==1)
          begin
            SLEEP_state <= SAVE;
            retain_n <= 1'b0;
          end
      end
      SAVE: SLEEP_state <= RESET_PDN;
      RESET_PDN: begin
        bius_pwr_reset_n <= 1'b0;
```

```
          SLEEP_state <= PWR_DOWN;
      end
      PWR_DOWN: begin
        pwr_dwn_req_n <= 1'b0;
        if (!pwr_dwn_ack_sync_n)
            SLEEP_state <= SLEEP_DONE;
      end
      SLEEP_DONE:
  endcase
endtask


//----------------------------------------------------
//                      wakeup_task
//----------------------------------------------------
task wakeup;
  case (WAKEUP_state)
      WAKEUP_IDLE: WAKEUP_state <= PWR_UP;
      PWR_UP: begin
        pwr_dwn_req_n <= 1'b1;
        if (pwr_dwn_ack_sync_n==1'b1)
            WAKEUP_state <= RESET_OFF;
      end
      RESET_OFF: begin
        bius_pwr_reset_n <= 1'b1;
        WAKEUP_state <= RESTORE;
      end
      RESTORE: begin
        retain_n <= 1'b1;
        WAKEUP_state <= CLOCKS_ON;
      end
      CLOCKS_ON: begin
        bius_gate_hclk_tmp <= 1'b0;
        h2pd_stop_pclk <= 1'b0;
        if (stop_pclk_ack_sync==0)
            WAKEUP_state <= CLAMPS_OFF;
      end
      CLAMPS_OFF: begin
        pwr_clamp_n_tmp <= 1'b1;
        bius_pwr_clamp_n_tmp <= 1'b1;
```

```
        h2pl_pwr_clamp_n_tmp <= 1'b1;
        bius_pwr_clamp_tmp <= 1'b0;
        if (pwr_clamp_ack_sync==0)
            WAKEUP_state <= WAKEUP_DONE;
    end
    WAKEUP_DONE:
  endcase
endtask
```

Figure 8-4 shows the top level state machine. The notation is the state-chart notation used in UML 2.0. We find this format more useful than the traditional bubble diagrams.

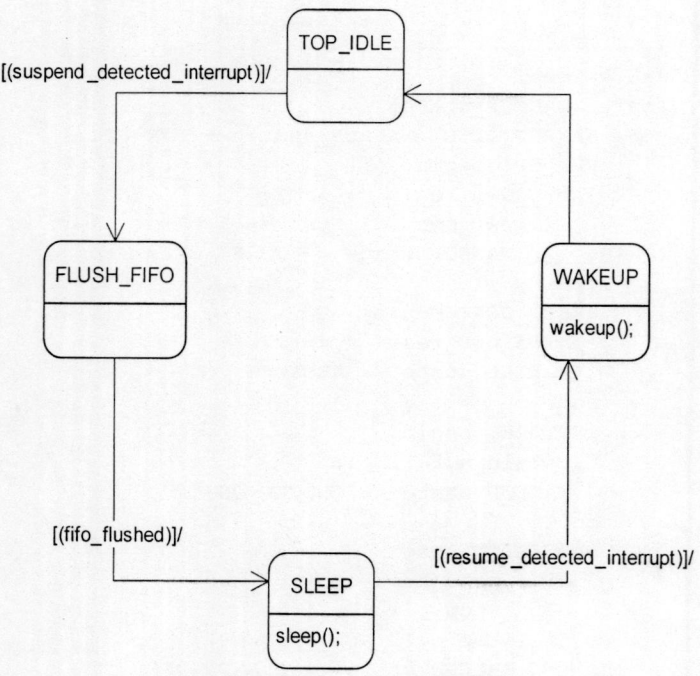

Figure 8-4   Top Level State Machine

The figure on the next page shows the details of the SLEEP and WAKEUP states:

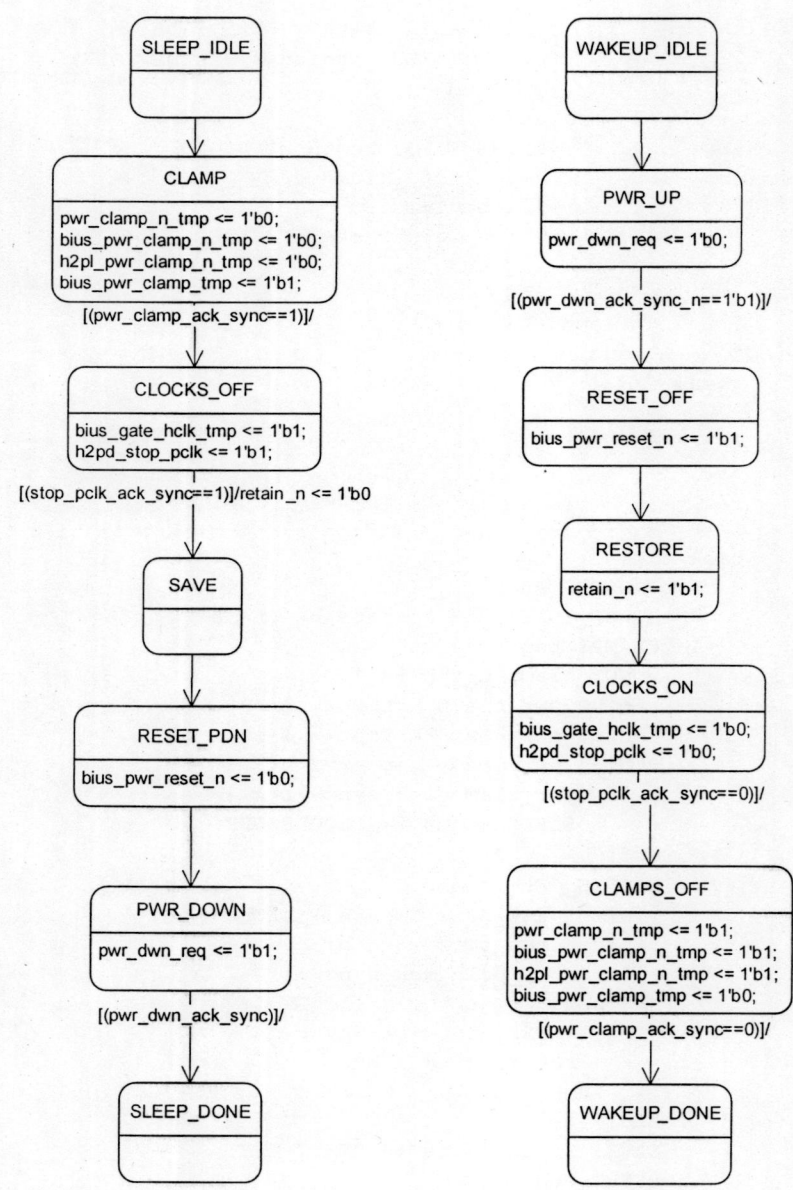

CHAPTER 9   *Frequency and Voltage Scaling Design*

Scaling the supply voltage of CMOS is possible over a technology-specific range; gate delays, setup and hold times and even memory access times scale monotonically with reduced operating voltage over a limited range. Linear voltage reduction results in a square-law reduction in both dynamic power consumption and in leakage power.

The earlier chapters have focused on basic multi-voltage techniques for optimizing dynamic power and on techniques to address leakage on advanced technology nodes. Voltage Scaling – reducing the supply voltage and clock frequency based on work load – is a more aggressive technique for dynamic power reduction. It can be effective on 0.18u and 0.13u technology nodes (typically 1.8 and 1.2V standard operating voltage respectively) where there is significant voltage headroom. In generic 90nm nodes (and below) there is not sufficient headroom to use voltage scaling very effectively. But it can be applicable to the "Low-Leakage" technology nodes at 90nm, 65nm and below, since these run at higher voltage than the equivalent generic or high-speed processes. (The 90nm low voltage processes run at 1.2V nominal voltage compared to 1.0V for the "generic" or high-speed process nodes, for example).

Voltage scaling introduces complications into both the system design and the implementation flow, but can be valuable for portable battery-powered products. Rarely is all the logic on a SOC required to run at the limit of performance at all times, and in many systems there may be several different performance profiles. Dynamically scaling the supply voltage to a processor or multi-media subsystem, for example, may significantly improve battery lifetime in the final product.

But every voltage scaled domain introduces another voltage regulator, usually off-chip, and the requirement to interface between different analog values across voltage boundaries.

In addition, determining the minimum voltage to meet a particular (sub)system performance level is a challenge. The EDA tools and library views start from a given set of process, voltage and temperature operating points, and sum delays across the design to determine critical paths and hence the resultant frequency limit. When the supply voltage of each block can vary among many different values, or even vary continuously, calculating delays and performing optimization becomes a much more difficult task.

The challenge of achieving timing closure over a range of voltages and clock speeds should not be underestimated. Even determining the appropriate range of voltages is a challenge.

The actual available headroom for a design is a complex function of process, design goals, libraries, and timing analysis methods. Temperature inversion in particular limits the range over which timing, voltage, and temperature maintain their normal monotonic relationship. A detailed analysis of all these factors is required before embarking on a DVFS design.

9.1   Dynamic Power and Energy

The dynamic power dissipated by CMOS is largely described by the equation:

$$P_{dyn} = C_{eff} \bullet V_{dd}^2 \bullet f_{clock}$$

Because dynamic power is linearly proportional to switching frequency, dynamically reducing the switching frequency whenever maximum performance is not required can reduce dynamic power significantly.

The fact that dynamic power is linearly proportional to the capacitance being switched is more of a design and implementation constraint and is improved primarily by reducing the length of interconnect driven and the design complexity and hence area.

The voltage term has the greatest effect on power, and in the case where frequency can be reduced to allow a reduction in voltage, the power is reduced quadratically.

Although generic in terms of technology and exact voltage, Figure 9-1 shows there is a region of operation where frequency increases monotonically over voltage, with a max voltage that is specified for the process, and a lower limit below which the circuitry runs out of safe voltage headroom and may fail to operate reliably – or where the delay paths no longer vary monotonically.

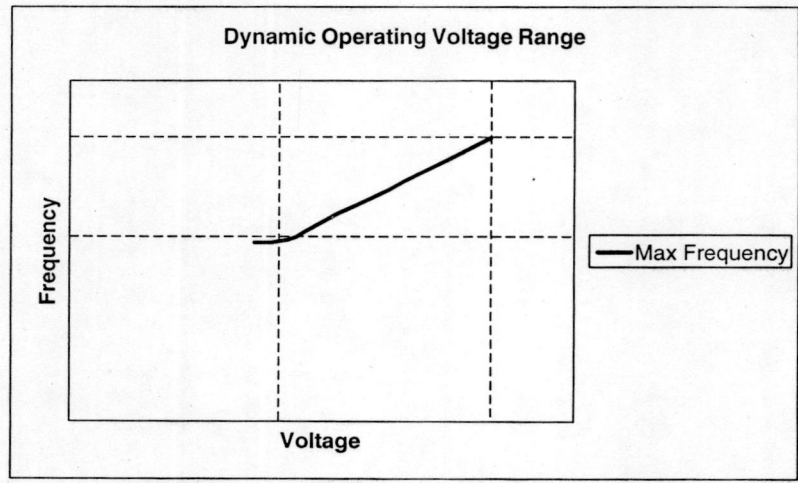

Figure 9-1   Voltage and Frequency Scaling Opportunity

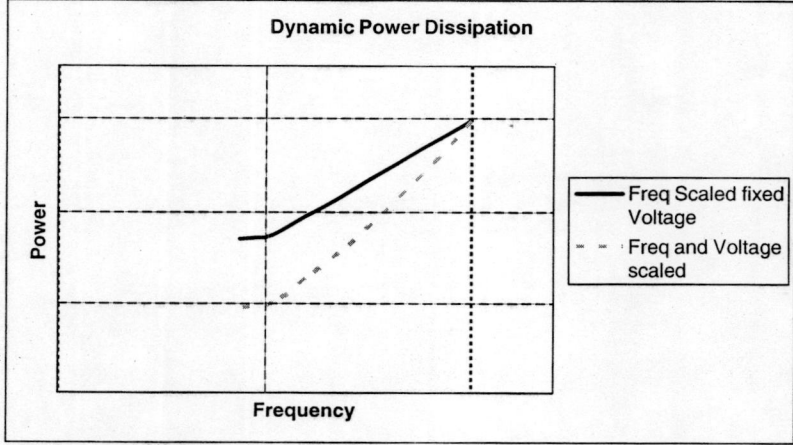

Figure 9-2   Increased Savings with Frequency Scaling

Figure 9-2 shows the generic power dissipation relationship between reducing frequency with and without reducing supply voltage. The gap between the two curves equals the power saving achievable between the minimum and maximum operating voltages.

Energy is the integration of power over the time taken to complete a task of work. Ignoring the effects of leakage power, clocking a block at half the frequency halves the dynamic power but takes twice as long to complete the work. Where scaling the voltage is possible the quadratic dynamic power reduction permits energy savings to accumulate over the duration of the task.

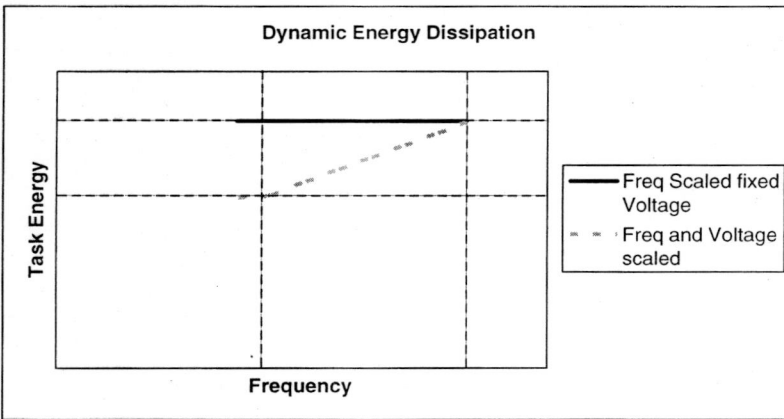

Figure 9-3   Energy Savings From Voltage Scaling

In calculating the energy savings from voltage scaling, the static leakage power cannot of course be ignored. Reducing the frequency and taking longer to complete a unit of work also means that the active leakage will be scaled in proportion to the inverse of frequency. Running to completion then stopping the clocks and applying the leakage mitigation techniques of the previous chapters allows the inactive leakage power to be minimized.

In addition, each voltage scaled block requires and additional power rail, and every power rail introduced into a SoC design has an impact on the realizable energy savings. Every regulated supply rail has some lost efficiency from generating that voltage with real world power controllers.

Although local on-chip voltage regulation would appear attractive, most digital CMOS technologies are not well suited to the implementation of switch-mode or linear voltage regulators that can power digital subsystems of more than a few tens or hundreds of gates.

9.2   Voltage Scaling Approaches

As stated in Chapter 3, the approaches to voltage scaling are:

* Static Voltage Scaling (SVS): different blocks or subsystems are given different, fixed supply voltages.
* Multi-level Voltage Scaling (MVS): an extension of the static voltage scaling case where a block or subsystem is switched between two or more voltage levels. Only a few, fixed, discrete levels are supported for different operating modes.
* Dynamic Voltage and Frequency Scaling (DVFS): an extension of MVS where a larger number of voltage levels are dynamically switched between to follow changing workloads.
* Adaptive Voltage Scaling (AVS): an extension of DVFS where a control loop is used to adjust the voltage.

In this chapter we focus on DVFS and AVS.

9.3   Dynamic Voltage and Frequency Scaling (DVFS)

Figure 9-4 on page 126 shows a simplified version of a DVFS design. The CPU subsystem is powered by a programmable power supply. The rest of the chip is powered by fixed power supply.

A PLL provides a high speed clock to the SysClock Generator, which uses dividers to generate the CPU CLOCK and the SOC CLOCK.

To execute voltage and frequency scaling, software first decides the minimum CPU clock speed that meets the workload requirements. It then determines the lowest supply voltage that will support that clock speed.

If the target clock frequency is higher than the current frequency, then the execution sequence is as follows:

* The CPU programs the power supply to the new voltage
* The CPU subsystem continues operating at the current clock frequency until the voltage settles to the new value
* The CPU then programs the new clock frequency.
* If the clock frequency change just requires a change in the divider value, it programs the SysClock Generator for this new value. No pause in CPU operation is required.

- If the new clock frequency requires a change in the PLL frequency, then the CPU programs the PLL to the new frequency. Either the PLL or the SysClock Generator suppresses all clocks until the PLL settles.

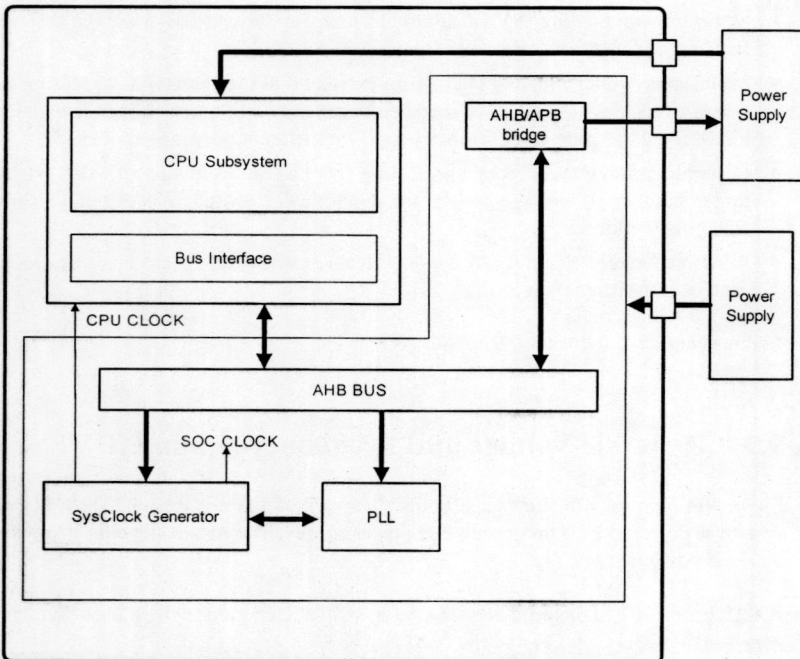

Figure 9-4   DVFS Block Diagram

If the target clock frequency is lower than the current frequency, then the execution sequence is as follows:

- The CPU first programs the new clock frequency.
- If the clock frequency change just requires a change in the divider value, it programs the SysClock Generator for this new value. No pause in CPU operation is required.
- If the new clock frequency requires a change in the PLL frequency, then the CPU programs the PLL to the new frequency. Either the PLL or the SysClock Generator suppresses all clocks until the PLL settles.
- The CPU then programs the power supply to the new voltage
- The CPU subsystem continues operating at the new clock frequency while the voltage settles to the new value

Varying clocks and voltages during operation is a new paradigm in design and offers some unique challenges:

- Determining which voltage and clock values to support
- Modeling timing
- Dealing with the settling time of clock generators and power supplies

Timing/Voltage Values

Most DVFS systems use a set of discrete voltage/frequency pairs. Determining which values to support is a key design decision, and is highly application dependent.

Too few operating points may result in systems that under some profiles spend a significant time ramping between the two levels – and the energy efficiency savings during the ramping times are typically significantly less than the steady-state values.

On the other hand, too many levels results may result in the power supply spending most of the time "hunting" between different target voltages.

Initially, we determine the number of operating points analytically:

- what are the appropriate clock frequencies for the different work loads
- which frequencies have clock periods that are multiples of the PLL period and thus require just changing the clock divider, not the PLL frequency
- what voltage is required to support each target frequency

Performing this timing analysis requires either special library and tool support or significant manual effort.

One approach to refining the selection of operating points is to provide the software developers with prototype, such as an FPGA implementation or a high level simulation model, that can run example workloads. There is no need to actually scale voltage; we simply emulate the performance clock scaling and representative power-supply ramp times. This enables us to understand how many distinct performance levels are useful under realistic dynamic workload conditions.

The Effects of Temperature Inversion

One restriction on DVFS design is that the voltages must be limited to the range over which delay and voltage track monotonically. That is, we must always operate above the temperature inversion point, the limiting voltage where delays start to behave non-monotonically with temperature.

Temperature inversion is an observed phenomenon on deep submicron technologies where delay and voltage invert their normal relationship. Normally, delay increases as temperature increases. But below a certain voltage ($2 * V_T$ according to [1]), this relationship inverts, and delay decreases as temperature increases.

Because this phenomenon is a function of V_T, the temperature inversion point varies between high V_T cells and low V_T cells. The result is that, if we decrease the voltage too much, we can see paths that used to be non-critical suddenly become the critical delay paths in our design. The voltage/timing relationship now becomes non-monotonic, and voltage scaling becomes (for all practical purposes) impossible.

Libraries

In order to establish what voltage levels are needed for the selected clock frequency, we need to perform timing analysis under a variety of conditions. Typically this involves doing some trial implementation at reduced voltages and measuring performance at these de-rated voltage points.

In order to do these trial implementations we need libraries whose characterization extends beyond their nominal supply voltages. Current source models, as described in Chapter 12, are an example of the type of models needed for this type of analysis.

Switching Times and Algorithms

For both voltage regulators and clock generators, switching performance levels takes time. We would like to the block to continue working (even at a reduced performance level) during this switching time.

Switching voltage levels is particularly slow. Off-chip linear regulators may take tens of microseconds to milliseconds to stabilize. Overshoot and undershoot need to be carefully controlled for safe DVFS operation, further increasing settling time. Typically we will want to multiple-source the voltage regulators, so the settling time may vary with the specific component used. The SOC designer needs to understand the power supply specifications in detail and how to handle multiple-sourced components with different DVFS characteristics – or build in worst-case delay counters to guarantee safe voltage settling times.

Switching frequencies is typically orders of magnitude faster than voltage level switching, especially if we just need to change the count value in the clock divider. (If we have to change the PLL frequency, then the worst-case PLL lock times may start to approach voltage settling times.)

We can take advantage of this faster clock switching time, but only with some care. Frequency can only be increased once the higher operating voltage has been safely reached. Frequency must be decreased before the dynamic scaled voltage is lowered.

Can the system make forward progress during DVFS changes? Stopping the clocks while a PLL re-locks may well be a requirement, but freezing clocks while waiting for voltage or the clock to change and settle could result in unacceptable interrupt servicing times – or break device driver timing requirements.

One technique to avoid this problem is to have the lowest operating frequency (such as a primary bus clock rate) available at all times. We can then use this clock while changing the supply voltage and the variable clock. This allows the block to continue safe operation as the voltage regulator and clock generator settle.

Power Up Sequencing

DVFS systems typically use at least two external power supplies. In this case we need to pay attention to the power-up control. We need to ensure that there are no deadlocks due to IO pad signaling not being stable until the power rail is valid, for example. We need to control the power up sequence and provide a guaranteed voltage settling time before issuing reset and starting the system. We can do this using a local digital counter, or some form of "voltage ready" handshake signal.

9.4   CPU Subsystem Design Issues

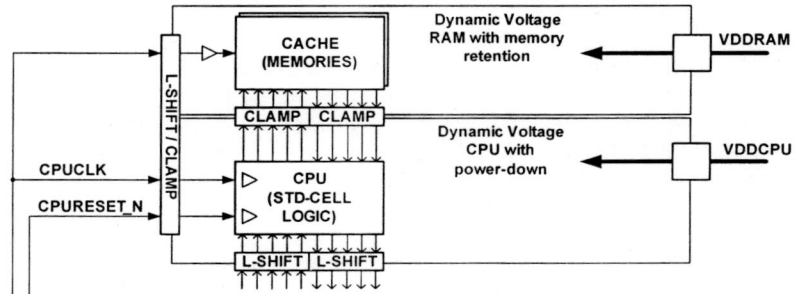

Figure 9-5   CPU Subsystem

DVFS is most frequently used on processor subsystems. Figure 9-5 shows an example of a cached CPU partitioned for voltage scaling and power gating. During power gating the CPU is powered down and the VDDRAM is set to the lower memory retention voltage. During voltage scaling the power supply is scaled to both the RAM and CPU logic domains together to ensure no differential voltage or timing across the critical path: the cache-CPU interface. In this case only isolation clamps and not level shifters are required across the CPU-memory interface. The clamps allow the cache memories to be isolated and held at a retention voltage rather than losing state during power down.

Level shifters are required between the CPU and rest of the chip. During power down the clock to the cache must be clamped as well. This means that there will be additional clock delay for the cache compared to the CPU. During clock tree synthesis we must compensate for this additional delay and achieve carefully balanced clock networks.

The partitioning described above is suitable for 130nm and above. Below 130nm there is little or no headroom for voltage scaling memories, so a more practical design is shown in Figure 9-6.

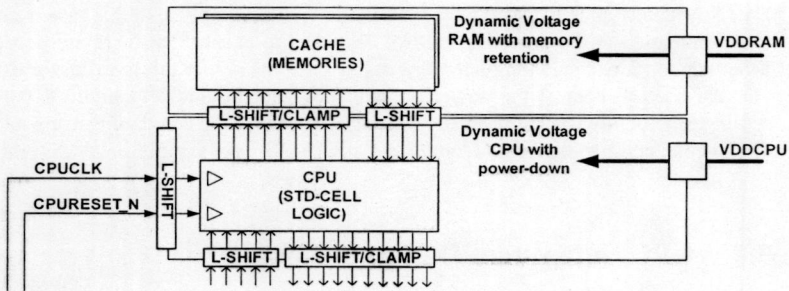

Figure 9-6   CPU Subsystem with Fixed Cache Supply

In this design, the cache uses a fixed, high voltage during operation. (During power down it can be set to a lower, retention supply voltage). Only the CPU is voltage scaled. Thus, we need to use level shifters between the CPU and cache, as well as on the other CPU interfaces. Also, all outputs of the CPU must have isolation clamps to support power gating.

Now that the CPU and cache run on different supply voltages, the clock frequency and latency for the cache memories must be scaled with the CPU supply voltage. In this case the clock for the cache is buffered with the standard clock tree in the VDDCPU domain to reflect the latency scaling and is then exported across the level shifter interface to the RAMs.

9.5   Adaptive Voltage Scaling (AVS)

The voltage scaling techniques described so far are "open-loop" techniques. Pairs of frequency/voltage values need to be determined with sufficient margin to guarantee

---

operation across the entire range of best and worst case silicon process and temperature.

In Adaptive Voltage Scaling a closed-loop feedback system is implemented between the voltage scaling power supply and delay-sensing performance monitor on the SoC. The on-chip performance monitor not only sees the actual voltage delivered on-chip but also understands whether the silicon is slow, typical or fast, and the effect of the temperature of the surrounding silicon.

Taking the cached CPU example again, the adaptive scaling tracking would be implemented with the voltage-scaled logic as shown in Figure 9-7.

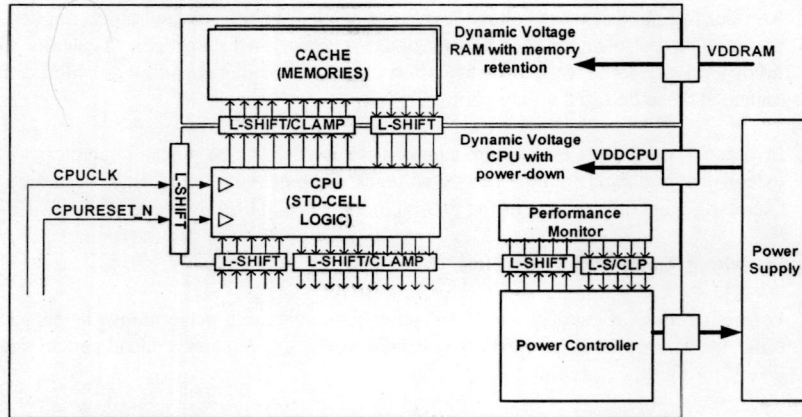

Figure 9-7   Adaptive Voltage Scaling

The performance monitor should be tightly integrated with the IP it is monitoring to get the best thermal tracking, and for a large voltage scaled subsystem there may be a number of distributed performance monitor blocks that can be analyzed together – with the worst sensor being the critical feedback element. The performance monitor communicates with a power controller which in turn set the voltage of the power supply.

9.6   Level Shifters and Isolation

As in any multi-voltage design, level shifters are required at the interfaces of blocks operating at different voltages. If the DVFS block is power gated, then we need to isolate the outputs as well.

Note that it is much easier to implement level shifters that shift only in one direction. That means that the DVFS block must always be at a higher voltage than the blocks it interfaces with or it must always be at a lower voltage.

Because of the lack of voltage headroom for RAMs, in most DVFS CPU designs, the cache is always at a voltage higher than or equal to that of the CPU.

Although in theory the bus interface of the CPU could be at a higher or lower voltage, for practical reasons the bus is usually also kept at a voltage higher than or equal to that of the CPU. The CPU can be carefully characterized to determine its minimum operating frequency; the bus interface unit usually is not so carefully characterized, and running it at or below the CPU's lowest voltage could cause system errors.

9.7   Voltage Scaling Interfaces – Effect on Synchronous Timing

The timing of a synchronous interface between a DVFS block and the rest of the system is made more complex by the fact that the DVFS block changes voltages and frequencies.

As the voltage in the DVFS block varies, so do the clock tree delays. There is no way to distribute a single, low-skew clock to both the DVFS block and the system that will remain low skew for all voltages. Thus, the standard model for a synchronous interface breaks down.

One solution is to use an asynchronous interface. One DVFS-enabled configuration of the ARM1176 takes this approach. It provides an asynchronous interface to an AXI bus, complete with synchronizers in both directions. These synchronizers do add to the initial access latency of the transactions across the interface. In this case, this increased latency is acceptable because the AXI bus is a split-transaction bus that can handle long-latency transactions without degrading the overall bus performance.

The more basic AMBA AHB bus does not support split transactions, and as a result long latency transactions directly degrade bus performance. Therefore, adding an asynchronous interface to an AMBA subsystem is not practical in most designs.

Figure 9-8 shows one approach to deal with this problem.

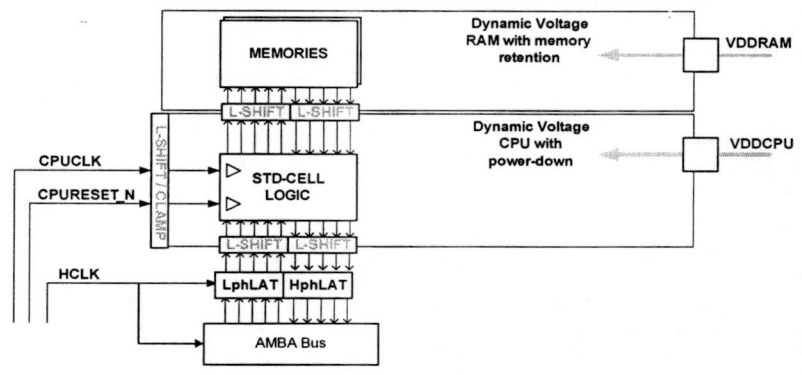

Figure 9-8   Latch-based Re-timing

This approach requires a CPU clock that is always a multiple of the bus clock (HCLK). We add latches at the interface between the CPU and the AMBA bus. The CPU clock is adjusted so that its rising edge occurs roughly aligned to the active (rising-edge) of bus clock HCLK With careful design, we can maintain this relationship to within half a CPU clock period over all operating conditions (including changing the voltage and clock frequency).

We then need to deal with the fact that the CPU clock can be early or late relative to HCLK. To deal with the case of an early CPU clock, we over-constrain synthesis to guarantee that data arrives early (by our worst case skew). If the CPU clock is late, the latch assures that data is still available.

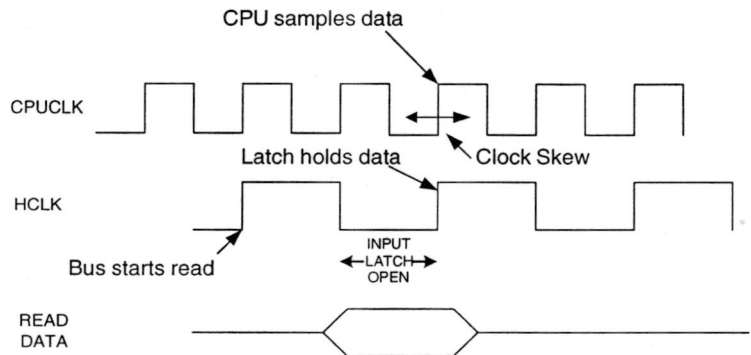

Figure 9-9   Read Timing for Latch Based Re-timing

The timing waveform for a read transaction is shown in Figure 9-9. In this example the CPU clock is twice the frequency of HCLK. The Low-phase input latches (LphLAT) are transparent when HCLK is low. Input data is guaranteed by over-constraining synthesis to arrive before the rising edge of HCLK. At the rising edge of HCLK, the latch captures the input data and holds the data for half an HCLK cycle. This guarantees that the data to the CPU will meet setup and hold requirements, even with significant skew on CPUCLK late relative to HCLK.

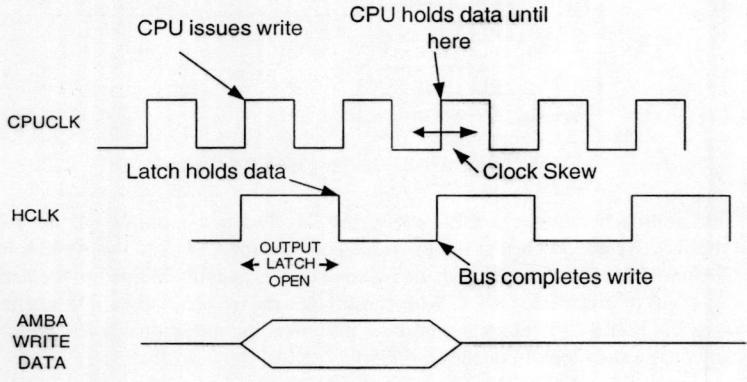

Figure 9-10   Write Timing for Latch-based Re-timing

Figure 9-10 shows how the output hold times to the SOC interface is managed. The High-phase HphLAT latches are transparent with HCLK is high. If the CPU clock is early, then the latch holds the old data on the bus until the write is complete. If the CPU clock is late, then data will be late arriving on the bus, so we over constraint the bus write timing in synthesis to guarantee that writes work correctly even if data is late by our worst case clock skew.

Thus for both read and write, the system level timing interfaces must be over-constrained to meet the worst case setup paths on inputs and outputs across the clock domains. The latch methodology assures that the hold times are managed safely in both directions.

An alternative approach to the CPU-Bus interface using standard rising-edge register is shown in Figure 9-11.

To avoid the complications of latches we run the CPU clock in advance of the bus interface clock, guaranteeing the hold times from the SOC to the CPU – providing the timing constraints on the SOC bus system are tightened to meet earlier timing to the CPU read path.

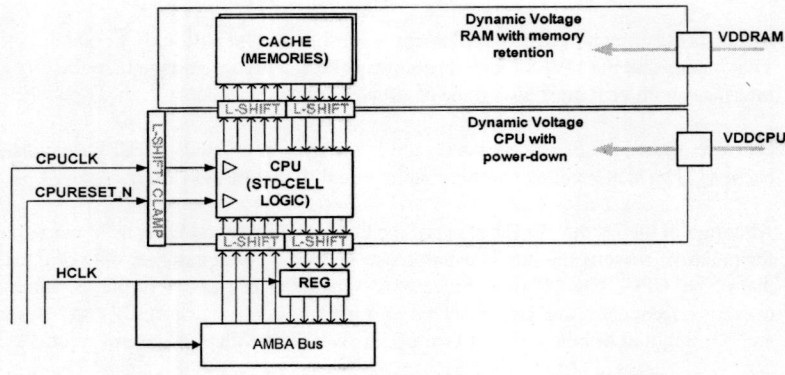

Figure 9-11   Register-based Re-timing

Figure 9-12 shows the write timing for this design. The CPU clock runs early enough so that it generates write data, the data is available at the input of REG early enough to meet the setup time requirements of the register. Write data is sampled by the register at the rising edge of HCLK and held for the duration of the write transaction.

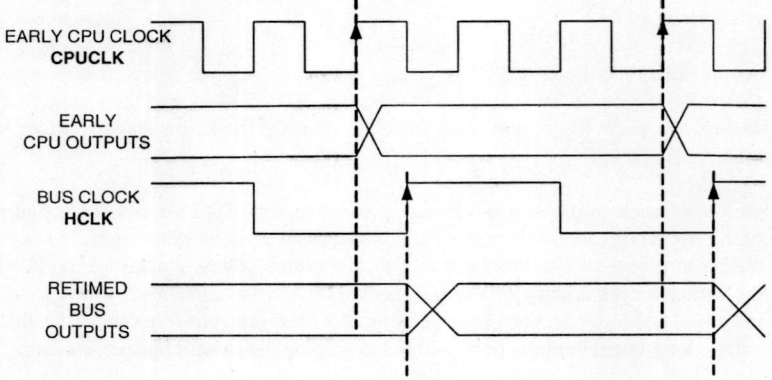

Figure 9-12   Write Timing for Register-based Re-timing

On read timing, we simply rely on the system to return read data before the CPU clock that occurs just before the rising edge of HCLK.

One advantage of this register-based approach is that it uses all edge-triggered registers, so standard implementation work effectively to assure correct timing. This makes automated design-for-test straightforward, but does requires tighter over-constraining of the input paths to the CPU interface.

9.8   Control of Voltage Scaling

Energy savings are only possible with Dynamic Voltage and Frequency Scaling when the system level performance requirements are understood and it is clear when the frequency can be lowered without missing deadlines or compromising the user interface "experience."

For an embedded system with a known workload it may well be possible to instrument the embedded firmware or hardware to drive the performance request and hence voltage requirements directly.

For a real-time system typically the deadlines are well understood and expressed in terms of scheduler priorities or scheduled events. These real-time requirements can be dynamically calculated and used to drive the performance and voltage scaling hardware. Because ramp-times for power supplies are not insignificant the delays for DVFS readiness can be factored into the scheduled deadlines in order to pre-compensate for such latencies.

Open application platforms are harder because typically these have to run downloaded tasks and applications that are not known (or may not even have been written) at system design time, and it is not feasible to require application writers to instrument their portable code to add SoC-specific DVFS hints or requirements in systems that will run a large number of concurrent applications.

Simply trying to guess from system utilization metrics or statistics is not a good solution to this problem. An example system utilization trace might look something like that shown in Figure 9-13. In order not to risk falling behind with sudden requests for performance, the system typically would have to keep erring on the side of higher performance. It could easily end up cycling between high performance and some "average" lower performance level, with some energy cost associated with needing to keep above the theoretical minimal level.

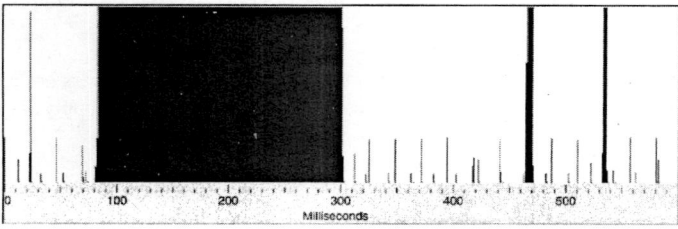

Figure 9-13   Example Utilization Trace

ARM's Intelligent Energy Manager (IEM) is an example of an approach that instruments certain operating system interfaces to build a task level view of the underlying requirements at the individual task or thread level. From this it builds an awareness of producer and consumer task frequencies and deadlines. For example, for a Graphical User Interface the calls to the window server and the perceived display refresh rates may be used to judge the right level of performance for interactive tasks.

An extensible stack of "IEM" policies is added alongside the operating system which then builds an aggregate level of performance sufficient to meet the deadlines of the dynamically changing task load.

The aim is to leave the user applications unchanged, although in a high-volume system design there may be additional energy savings possible from simple annotation of key real-time tasks to indicate their specific requirements to the policy stack.

Figure 9-14 shows how the example utilization trace shown in the previous figure can be broken down with task knowledge to determine that certain tasks are periodic. In this particular example there happened to be a lightweight sound daemon running frequently. This daemon is interrupt driven and consumes data from a main audio decoder producer task. This task needs to produce the next buffer of sound samples every 180ms or so. In the mean time, the foreground user task consumes the rest of the bandwidth some of the time, or is waiting on user interaction much of the rest of the time.

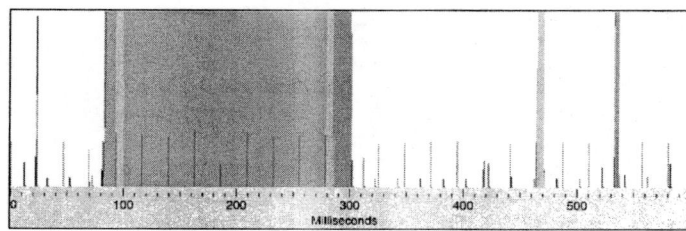

Figure 9-14   Trace With with Periodic Tasks Identified

Whatever the underlying system design or software workloads, the energy savings will only be as good as the decision making algorithms that set the performance level.

References

1. Lasbouygues, B. et al.,"Temperature Dependence in Low Power CMOS UDSM Process", in E. Macii et al. (Eds.) *PATMOS 2004*, LNCS 3254, 2004.

CHAPTER 10   *Examples of Voltage and Frequency Scaling Design*

The previous chapter introduced Dynamic Voltage Scaling and Adaptive Voltage Scaling. This chapter describes two examples of voltage scaling: the ULTRA926 and the ATLAS926. Both chips were designed as technology demonstrators for low power design techniques.

10.1 Voltage Scaling - A Worked Example for UMC 130nm

The ULTRA926 technology demonstrator chip used an ARM926EJ-S as a test vehicle for Adaptive Voltage Scaling. The ULTRA926 was a collaborative effort between ARM, Synopsys, Artisan, NSC, and UMC. The overall SoC design was developed by ARM for UMC with Synopsys providing the EDA methodology and implementation, Artisan providing the IEM-specific cell library and PLL components, and National Semiconductor Corporation providing Adaptive DVFS (AVS) power supply technology.

The chip was implemented using UMC's 130nm process. This process has a nominal supply rail voltage of 1.2V, which leaves some useful headroom to apply DVFS techniques and obtain valuable energy savings.

As a methodology development vehicle this presents the primary design and verification problems that have to be addressed for full-scale product designs.

10.1.1 ULTRA926 System Design Block Diagram

As shown in Figure 10-1the ULTRA926 system supports both DVFS and AVS.

The cache memories and CPU standard cell logic share the VDDCPU power domain as there was sufficient headroom to voltage scale the memories in this technology. So they are voltage scaled and power gated together.

The tightly coupled memories (TCMs) have their own voltage-scaled domain, VDDRAM, which is scaled with the VDDCPU voltage in functional use, but can also be kept alive at a reduced retention voltage when the CPU is powered down. The clamps allow the RAM control signals to be isolated when the CPU is powered down.

Level shifters handle the voltage scaled interface between the CPU subsystem and the SoC. The CPU clock is generated in advance of the AHB clock to allow the simple HCLK re-timing registers on CPU outputs to the SOC.

The SoC power domain contains the PLL's, power management, clock generation, system memory and peripheral controllers. The SoC domain is always powered up.

The CPU subsystem was implemented on the UMC 130HS process technology which is faster but leakier than the 130LL Low Leakage technology used to implement the rest of the SOC.

Support for Adaptive Voltage Scaling is included for the VDDCPU domain. The Adaptive Power Control IP is implemented in the always-on VDDSOC power domain together with the PLL(s) and the clock generator.

For the adaptive voltage scaling, the current clock (guaranteed safe CPU frequency for operation at current voltage conditions) and a target clock (desired target CPU clock frequency) are generated. On the request to raise the performance level, the CPU continues to be clocked at the safe current clock frequency while the voltage on the VDDCPU domain is raised. When the Hardware Performance Monitor has sufficient timing slack to indicate that the voltage is high enough, the Dynamic Clock Generator (DCG) is switched up to the target frequency.

On the request to lower frequency, the DCG provides the new, lower clock to both the CPU and the Hardware Performance Monitor. The power supply is then ramped down to just meet the required timing slack and margins for the reduced target frequency.

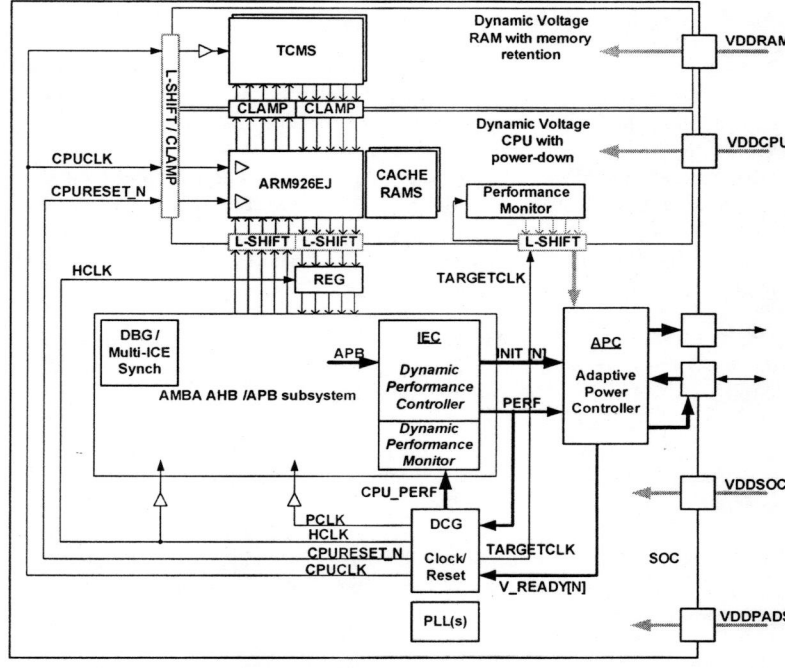

Figure 10-1   ULTRA926

10.1.2 Voltage/Frequency Range Exploration

Selecting appropriate frequencies for DVFS on the ULTRA926 project required a good understanding of the voltage/delay characteristic of the system. Unfortunately this is a compound surface as the critical paths are a mix of combinatorial, sequential and memory components. Typically one must start from a number of voltage points and determine the composite delays in order to derive the frequency of operation.

On this project we used transistor level models to analyze the performance of the core at reduce voltages as cell based timing models were not considered to be accurate when de-rated beyond standard IR drop levels. A Synopsys NanoSim characterization flow was used to generate detailed analysis of all paths and post-processed to generate top level timing models of the core in Liberty format (.LIB)

As this process is highly compute intensive it was not practical to analyze the core at fine grain voltage decrements across process and temperature. However the voltage-delay characteristics for buffers, combinatorial cells, sequential cells and memory elements has been analyzed by others and the monotonicity for each has been confirmed and well documented [1].

The standard CPU characterization at 1.2 (nom), 1.08 (-10%), 1.32 (+10%) for fixed voltage rail operation and IR drop was performed for the fast and slow process and temperature corners. From experience and knowledge of the RAM operating voltage headroom, the characterization was extended down to 0.94V, 0.80V and 0.73V operating points. Some further analysis was in fact conducted down at 0.66V but the accuracy of the RAM models was of concern and not trusted for detailed design work.

The resulting maximum frequencies at the characterized voltages are shown plotted in Figure 10-2.

The standard "sign off" worst case frequency (Fmax [SS] at 1.08V) was 288MHz. As the RAM sense amps only need approximately 0.75V, this frequency could be comfortably halved to 144MHz as our minimum DVFS performance point. Analysis of a previous DVFS implementation showed that the best energy savings are in the 50-100% frequency range, so bounding the clock by 288MHz (max) and 144MHz (min) made sense. The challenge was to determine the intermediate frequencies and how to generate them.

A PLL master clock operating at 2 x 288MHz (576MHz) provided the starting point. To provide fast clean dynamic clock switching, we used a shift-register approach to generate the CPU clocks shown in Table 10-1:

Adding the option of a second PLL locked at 5/6 of the main PLL (480MHz) provides the 240MHz frequency. The power costs – and potential dynamic power-down – of a second PLL of course have to be factored into the system level energy savings and are highly technology dependent.

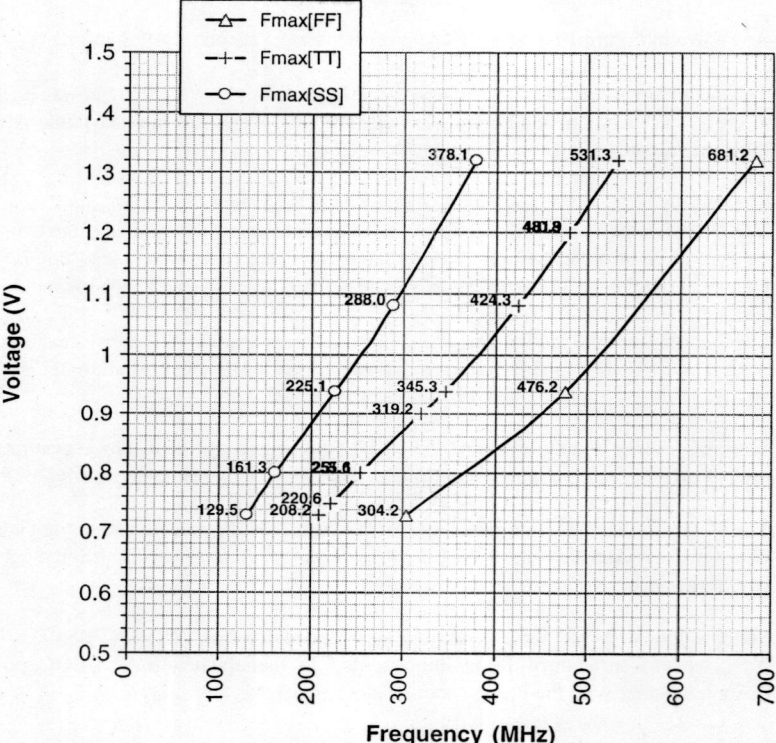

Figure 10-2   ULTRA926 Scaling Analysis

Table 10-1

PLL clock (MHz)	Divider	CPU Clock (MHz)	CPU Performance %	Voltage (Worst) (V)	Voltage (Typ) (V)
576	2	288	100%	1.08	0.87
480	2	240	83%	0.95	0.8
576	3	192	67%	0.86	<0.75
576	4	144	50%	0.76	<0.75

10.1.3 Synchronous Design Constraints

Varying the voltage results not only in the variable performance (and set up and hold times) but also in wide clock tree latency variation as shown in Figure 10-3:

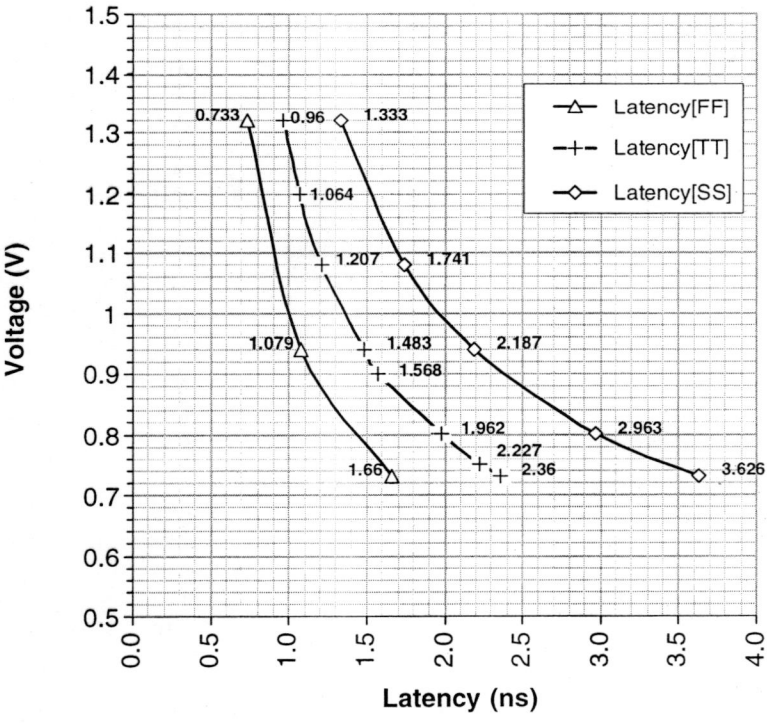

Clock Tree Latency Analysis

Figure 10-3 ULTRA926 Clock Tree Latency

Such a wide variation of latency is impossible to handle with standard set-up and hold fixing, and so the interface to the core looks to be asynchronous. However, it was possible to avoid the overhead of synchronizers by "pre-compensating" the clock to accommodate the increased latency in the core. This technique also requires a set of "re-timing" registers ("REG" in the block diagram) on the outputs to ensure hold times are met.

10.1.4 Simulated (predicted) Energy Savings Analysis

The dynamic power component is governed by the $\{CV^2f\}$ switching component for CMOS. With fixed voltage, executing at half speed effectively halves the dynamic power – but takes twice as long to complete the workload. The total energy consumption is given by the product of power and duration (reciprocal of the performance) to complete the same workload so the energy calculation is proportional to the V^2 component. The relative energy efficiency for the candidate frequencies is shown in Table 10-2:

Table 10-2

CPU Clock MHz	Voltage (worst) (V)	Voltage (typ) (V)	V^2 (worst) (E)	V^2 (typ) (E)	Energy Ratio (worst)	Energy Ratio (typ)
288	1.08	0.87	1.166	0.757	100%	65%
240	0.95	0.80	0.903	0.640	77%	55%
192	0.86	0.75	0.740	0.563	63%	48%
144	0.76	0.75	0.578	0.563	50%	48%

For worst-case silicon there is on the order of 23%, 37%, and 50% energy savings when one can run the tasks at selectively lower performance levels. However in the normal production spread the typical silicon at room temperature is close to 35% more energy efficient at maximum performance and tends to just better than 50% savings for the lowest two performance levels. Interestingly with typical silicon the 50% performance level (144 MHz) is no more energy efficient than 67% (192 MHz) as there is no voltage scaling headroom left to exploit – an important consideration in designing production silicon for optimal battery-life.

Note. All the analysis has been performed at the lower limit of operation. Safe working tolerances do need to be added for system-level IR drop and operating margins. The relative analysis for an additional 10% (etc.) tolerance simply affects the overall magnitude and not the energy ratios between operating points.

10.1.5 Silicon-Measured Power and Performance Analysis

The measured current I/V and power P/V curves are included below, and show good monotonicity. The measurements are at the power supply connections to the board power connector. Thus, the actual silicon has additional IR drop from the circuit board, BGA socket, bonding wires and on-chip power rails – while the transistor simulations do not take any of these into account. So the fact that we could maintain operation down to 0.72V with typical silicon at room temperature was encouraging.

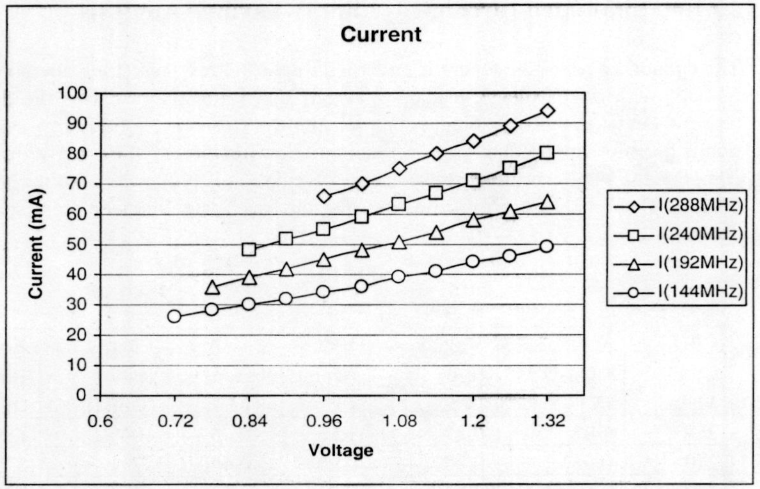

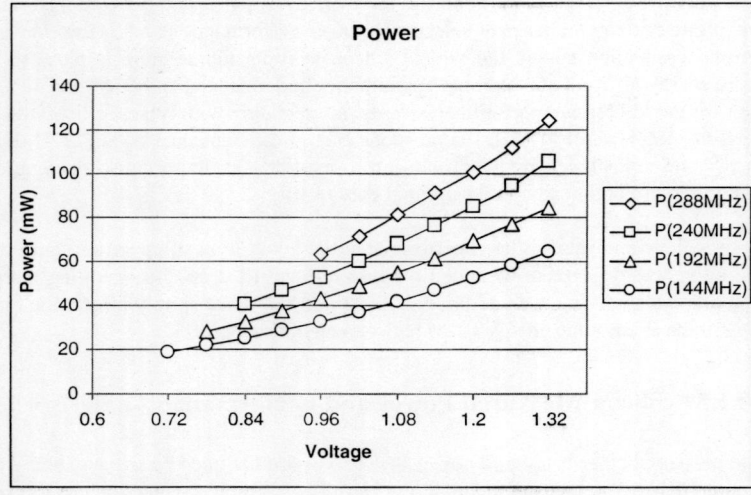

Figure 10-4   Current, Voltage and Frequency for ULTRA926

Power

Figure 10-5   Power, Voltage and Frequency for ULTRA926

The processing performance for the cache-intensive Dhrystone application work load and how this translates into work-load duration for the energy consumption analysis is tabulated below:

Freq (MHz)	KDhry/ second
144	217.00
192	288.32
240	360.58
288	432.91

Freq (MHz)	1MDhry(milliseconds)
144	4608.30
192	3468.31
240	2773.30
288	2309.97

10.1.6   Silicon-Measured ULTRA926 DVFS Energy Savings Analysis

Table 10-3 on page 148 shows the tabulated energy (power * duration) measurement data displayed at the four performance levels designed into the ULTRA926 SoC. These were gathered across 5% steps of VDDCPU supply from 60% to 110% of the nominal (1.2V) supply rail. A blank in the column entry indicates the processor is outside safe operating range.

The frequencies chosen for the ULTRA926 project are in fact all multiples of a master 48MHz bus clock at the default PLL multiplier configuration setting:

- 6x (100%) for worst case 288MHz "FMax" sign-off
- 5x (83.3%) for 240 MHz
- 4x (66.7%) for 192MHz
- 3x (50%) for 144MHz

The energy measurements all correlate very cleanly – taking a row for a certain operating voltage, the energy consumption is similar at each of the frequencies with some positive increase approaching 5% in the 144MHz case.

Table 10-3   Energy vs. Operating Voltage

V	288MHz (mJ)	240MHz (mJ)	192MHz (mJ)	144MHz (mJ)
0.72	(unsafe)	(unsafe)	(unsafe)	37.44
0.78	(unsafe)	(unsafe)	42.12	43.68
0.84	(unsafe)	48.38	49.14	50.40
0.90	(unsafe)	56.16	56.70	57.60
0.96	63.36	63.36	64.80	65.28
1.02	71.40	72.22	73.44	73.44
1.08	81.00	81.65	82.62	84.24
1.14	91.20	91.66	92.34	93.48
1.20	100.80	102.24	104.40	105.60
1.26	112.14	113.40	115.29	115.92
1.32	124.08	126.72	126.72	129.36

Figure 10-6 shows the same energy consumption data plotted in histogram form to visually display the close-to-linear energy efficiency relationship measured for the device.

Even with 10% voltage margins added back for safety there is still on the order of 50% energy savings on workloads that can be run for twice as long at half the frequency (50.4 milli-Joules for 144MHz at 0.84V compared to 100.8 milli-Joules for 288MHz at 1.2V).

The leakage power becomes apparent in the energy "losses" for the 83%, 66% and 50% performance levels compared to 100%. If we look at any one row of the table (1.20V, for example) we see that the energy increases as clock speed decreases. Thus, even though the dynamic energy is the same for each of the four cases, the leakage current increases the total energy for lower clock speeds running longer. For leakier process technology nodes this is a reminder of the balance that must be evaluated between running slower to reduce power and the expense of burning leakage power for longer.

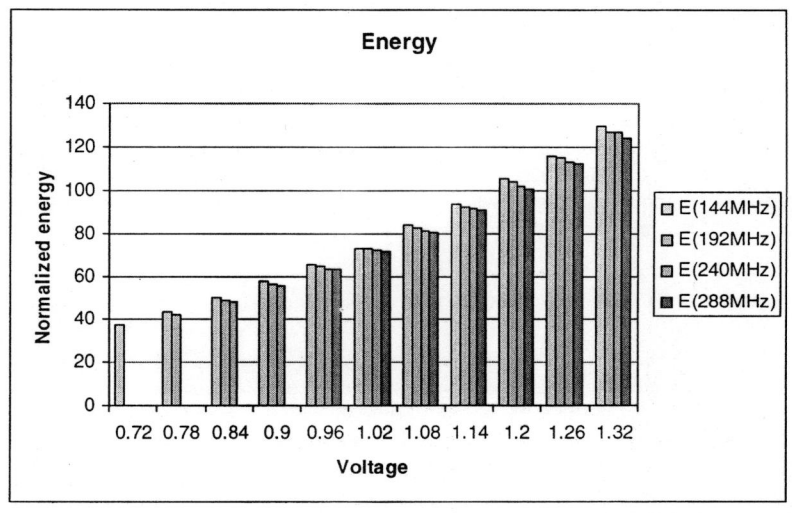

Figure 10-6   Energy, Voltage and Frequency for ULTRA926

The "shuttle" silicon for this project was confirmed as close to typical process by UMC. To explore the case of slower silicon, we raised the master PLL frequency to raise FMax to 360MHz.

The measured minimum voltages for the four supported fractional performance levels are tabulated below together with the energy consumption savings possible while emulating "moving the silicon" closer to the edge by using over-clocking:

CPU MHz	Vmin (limit)	I (mA)	KDhry/sec	Energy (mJ)	Energy consumed
180	0.777	33	271.003	95	58%
240	0.842	47	363.636	109	67%
300	0.932	65	454.546	133	82%
360	1.030	86	542.005	163	100%

10.2   Voltage Scaling – A Worked Example for TSMC 65nm

At 65nm, process technologies diverge into higher performance (but higher leakage power) or lower leakage versions with different gate oxide materials. The "generic" process technologies are typically 1.0V nominal supply voltage while the low-leakage variants use 1.2V nominal supply voltages. The low leakage technologies can suffer from higher dynamic power when pushed for performance so voltage scaling is a reasonable option for optimizing battery-life in this technology.

10.2.1   ATLAS926 Case Study

The ATLAS926 project was a collaborative effort between ARM and TSMC to demonstrate both dynamic and static power reduction on a low leakage 65nm technology. This "LP" process is a 1.2V nominal technology that is optimized for portable products, and contrasts with the higher performance "generic" technology that is specified at 1.0V and exhibits a higher leakage power.

Although the 1.2V LP process is low leakage, the dynamic power dissipation needs careful management when high clock speeds are required. (The quadratic dependence on voltage implies that moving from 1.0V to 1.2V increases dynamic power 44%.)

The compiled RAM technology available to this project however was not safely scalable with voltage. The full 1.08V (1.2V – 10%) worst case voltage is required for safe RAM operation. Therefore low-leakage High-V_T RAM technology was specified for the cache memories to keep the retention power to a minimum when not running, and voltage scaling only applied to the standard cell logic.

The standard cell libraries and level shifters available to the project had extended low voltage characterization in addition to the standard +/- 10% derating from nominal:

Table 10-4   Voltage Range for Library Characterization

Voltage Scaling	Voltage (V)
110%	1.32
100%	1.2
90%	1.08
80%	0.96
70%	0.84

Because the cache RAMs were not voltage scalable, the level shifters had to be introduced into the ARM926 CPU design to handle the different VDDCPU and VDDRAM voltage interface levels. As has been described earlier, low-to-high level shifters in particular add some delay. Given the fact that critical paths in a cached microprocessor are often across the memory interfaces, this did introduce complexity and limit the upper frequency of operation at worst case conditions.

10.2.2   Voltage/Frequency Range Exploration

The design approach for this project was to establish the worst-case performance for standard condition sign-off (the worst case temperature and process conditions at 1.08V). With the level shifters on the RAM and SOC interfaces this was found to be around 240 MHz. With careful implementation work this was improved to 250MHz, but the cost/timescale trade-offs were clearly in the area of diminishing returns.

Unlike the ULTRA926 project there was not the time available to do weeks of detailed transistor level simulation so the approach was to design the clock generator to support a range of instantaneously switchable frequencies and determine the safe operating voltage and margins later.

The dynamic clock generator was specified with a 1GHz master PLL clock that supported integer divider ratios for the CPU, system bus and memory clocks, in order to support a synchronous interface between the cached ARM CPU and the AHB system bus IP.

Frequency (MHz)	Performance Level
250	100%
200	80%
150	60%
100	40%
50	20%

10.2.3   Silicon-Measured Power and Performance Analysis

The VDDCPU and VDDRAM supplies were separately bonded out of the ATLAS design in order to support independent current monitoring and safe working voltage testing.

The limits of voltage scaling were mapped and the current measurements for reliable operation were captured. These were used to derive the measured steady-state power graphs shown below.

The graphs are basically monotonic but the current resolution of 1mA resulted in slightly quantized current readings resulting in the staircase appearance of the graphs. Below 1.08V a shallower gradient is just apparent. Down to 1.08V the CPU and cache were scaled together; below 1.08V, only the CPU was scaled. This resulted in the partial flattening of the curve.

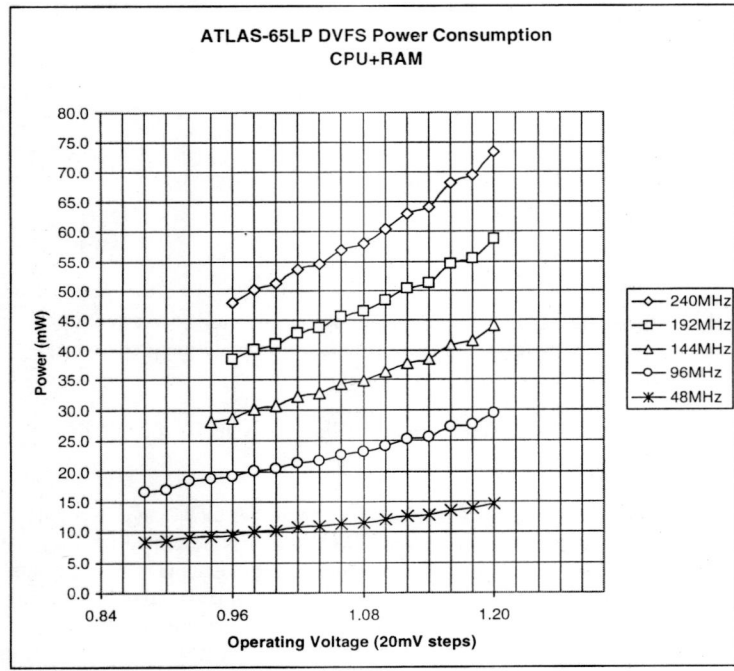

Figure 10-7   Power, Voltage and Frequency for ATLAS926

Figure 10-8 documents the measured energy consumption after the workload duration scaling has been factored in. The energy is shown normalized to the consumption at 1.20V. Again the energy efficiency gradient is slightly shallower below 1.08V due to the RAM energy cost remaining consistent below this point.

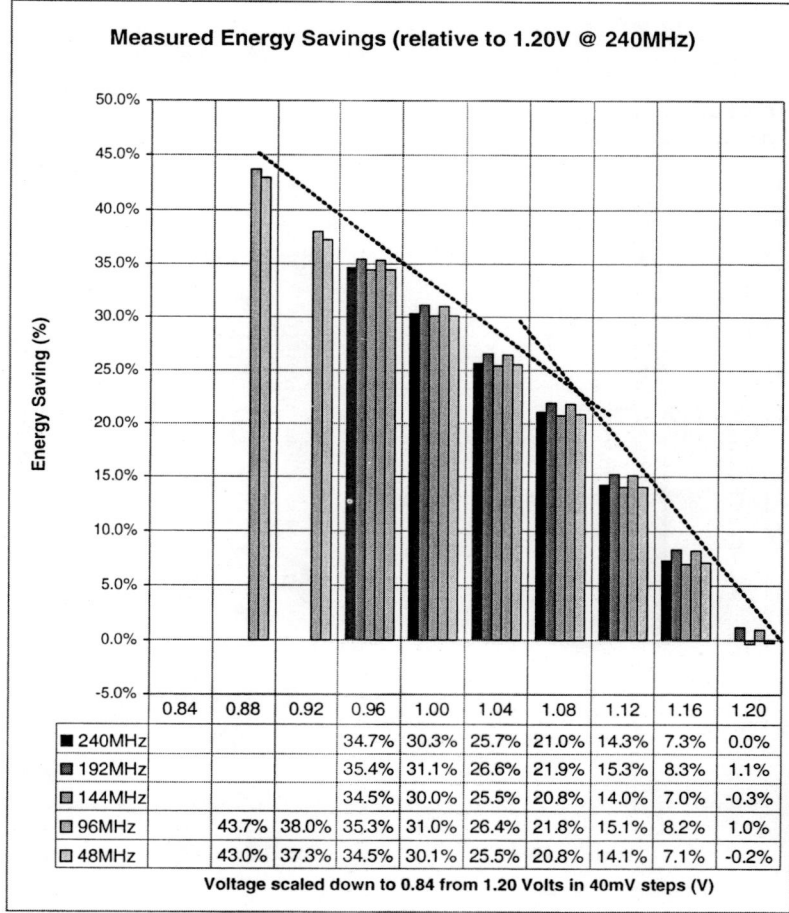

	0.84	0.88	0.92	0.96	1.00	1.04	1.08	1.12	1.16	1.20
■ 240MHz				34.7%	30.3%	25.7%	21.0%	14.3%	7.3%	0.0%
▨ 192MHz				35.4%	31.1%	26.6%	21.9%	15.3%	8.3%	1.1%
▤ 144MHz				34.5%	30.0%	25.5%	20.8%	14.0%	7.0%	-0.3%
▨ 96MHz		43.7%	38.0%	35.3%	31.0%	26.4%	21.8%	15.1%	8.2%	1.0%
▯ 48MHz		43.0%	37.3%	34.5%	30.1%	25.5%	20.8%	14.1%	7.1%	-0.2%

Voltage scaled down to 0.84 from 1.20 Volts in 40mV steps (V)

Figure 10-8   Total Energy (Relative to 1.20V FMAX)

Dynamic voltage scaling also has a dramatic effect on leakage power in as well as dynamic power. Figure 10-9 on page 154 shows the measured leakage power at room temperature for the ATLAS silicon, plotted on a logarithmic scale to handle the dynamic range:

- The HALT curve is the baseline leakage when clocks are stopped.

- The RETENTION curve is the measured leakage with State-Retention Power-Gating

- The HIBERNATE curve is the measured leakage with the power supply to the standard cell logic turned off. Thus it is a measure of the cache leakage.

The breakpoint in the curves at 1.08V reflects the fact that only the logic portion of the CPU is scaled below this voltage; the RAM supply rail is not scaled below 1.08V.

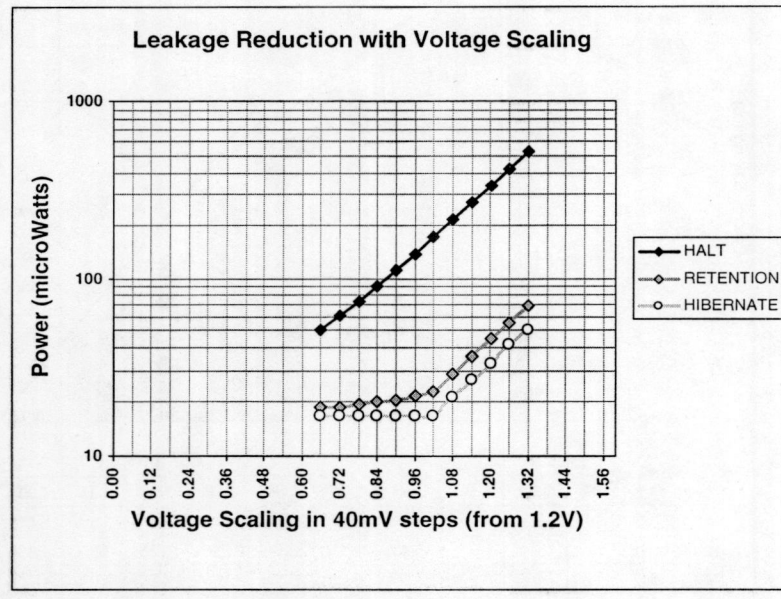

Figure 10-9   Power Savings in the Different Modes

References

1. Thomas D. Burd, Trevor Pering, Anthony Stratakos, Robert W. Brodersen, "A Dynamic Voltage Scaled Microprocessor System," 2000 IEEE, ISSCC 2000, paper 17.4.

CHAPTER 11

Implementing Multi-Voltage, Power Gated Designs

This chapter describes the implementation of designs that use power gating and multi-voltage techniques. It highlights the areas in the implementation process that are specific to multi-voltage and power gating designs.

To illustrate the implementation process, we use a multi-voltage power gated ARM1176JZF-S microprocessor as an example design.

The ARM1176JZF-S integrates a number of technologies including power gating and dynamic voltage scaling. It employs the IEM techniques described in Chapter 9; it dynamically monitors and predicts the performance requirements of multiple applications, and tunes the processor's operating voltage and frequency to match the requirements. These techniques reduce the processor's energy consumption by 25% - 50%. In order to exploit this IEM technology, the ARM1176JZF-S processor has been architected for a low power implementation.

The ARM1176JZF-S is illustrated in Figure 11-1 on page 156 and consists of a cache sub-system, central core CPU, memory management sub-system and AXI interface.

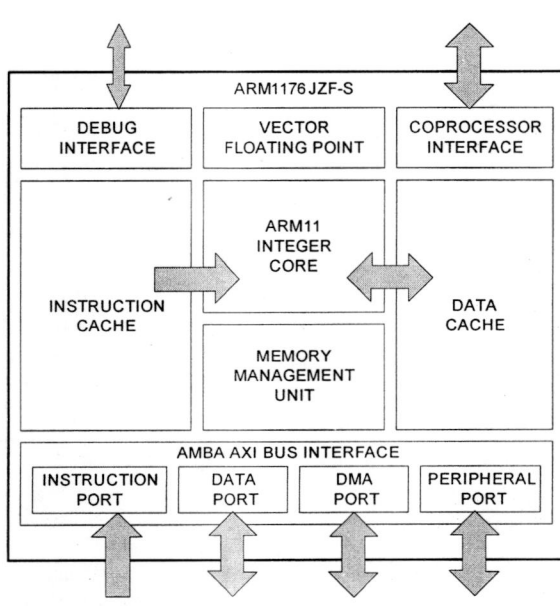

Figure 11-1   ARM1176JZF-S Synthesizable Applications Processor

There are three power domains within the ARM1176JZF-S

- An always-on power domain (VSOC) containing the logic that interfaces to the SoC through the AXI interface. The logic in this power domain also manages the asynchronous mode of the processor when performance scaling is employed.

- A shutdown power domain (VCPU) that contains all of the core CPU logic. This power domain can operate at multiple voltages and can also be powered down.

- An always-on power domain (VRAM) that contains the cache memory instances. Typically, memory cannot be scaled to the same degree as standard cells and during voltage scaling the VRAM and VCPU power domains need to be scaled independently. Also, when VCPU is shutdown, the VRAM power domain can be placed into retention. However, as discussed earlier in this book, if the cache is placed into retention state during sleep then the wake-up from sleep will take longer since the time for the power supply to the memories to stabilize will need to be factored into the wake-up time.

Figure 11-2 below shows the basic structure of the ARM1176JZF-S from a power domain perspective.

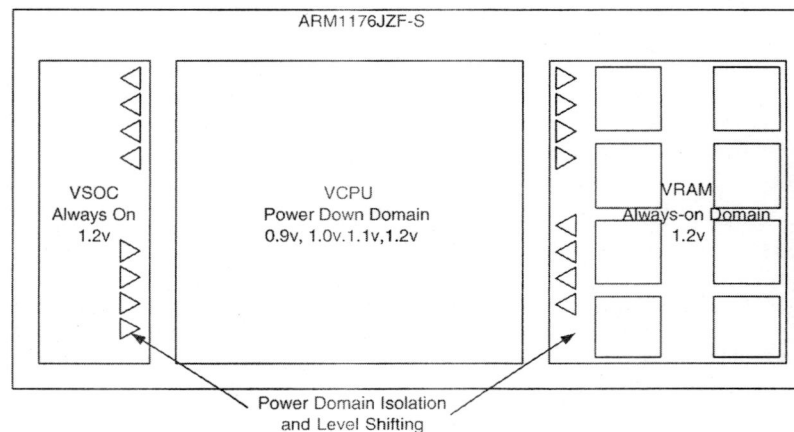

Figure 11-2   ARM1176JZF-S Multi-voltage, power gated design

VSOC always operates at full voltage – the same voltage as the logic sitting on the AXI bus externally to the processor. The VCPU power domain can operate at a number of voltages and can be shutdown. VRAM operates at the same voltage as VSOC or is placed into retention if VCPU is shutdown. Placing a memory into its retention state typically means stopping the clock(s), disabling the memory and reducing its supply voltage to the retention voltage.

To implement this design, we use an implementation process that allows for the logical and physical partitioning of multiple power domains with multiple voltage levels. The implementation process also supports the insertion of level shifters, isolation cells, and retention flops. We will also provide power switching networks for the VCPU power domain. As we proceed with the implementation, we pay particular attention to management of voltage drop, impacts on timing from our multi-voltage partitioning and the electrical integrity of our design, particularly at the power domain interfaces.

For the discussion in the rest of this chapter, we introduce two definitions:

A *Power Domain* is a logical entity and is a collection of design elements that share a primary power supply.

A *Voltage Area* is a physical entity that represents a geographic area of the chip that is used to place the logic within a given power domain. It is possible for a power domain to have a number of disjoint voltage areas depending on the nature of the chip. However, in the discussion that follows, it is assumed that there is a one to one relationship between a voltage area and a power domain.

Power domains are created during the synthesis phase; the voltage areas (physical realizations of the power domains) are created during design planning and managed throughout the rest of the flow. Levels shifters, isolation cells and retentions registers are inserted into the design as early as possible so that their impacts to both timing and the physical design can be considered. Clock tree synthesis and design for test must also be power aware.

11.1 Design Partitioning

Partitioning a design into separate power domains introduces new interfaces between each power domain. These added interfaces may contain isolation cells and level shifters. Thus, the specific partitioning can have a significant impact on the overall performance of the design.

The job of partitioning a design into power domains is a joint responsibility of the system architect(s), RTL designer(s), and implementation engineer(s). Between them they understand the target application, the power and performance objectives, and the limitations of the target technology.

11.1.1 Logical and Physical Hierarchy

When mapping the logical hierarchy to power domains, we want to assign hierarchical functional units in their entirety wherever possible. Synthesis engines operate top down on a cell in the hierarchy and the best Quality of Results (QoR) is achieved if the entire hierarchy can be manipulated under the same goals and constraints.

Gathering up various hierarchical cells from different levels of the design and assigning them to power domains can make timing closure significantly more difficult.

Figure 11-3 shows the logical hierarchy of the ARM1176JZF-S after some logical grouping of top level modules in preparation for a low power implementation.

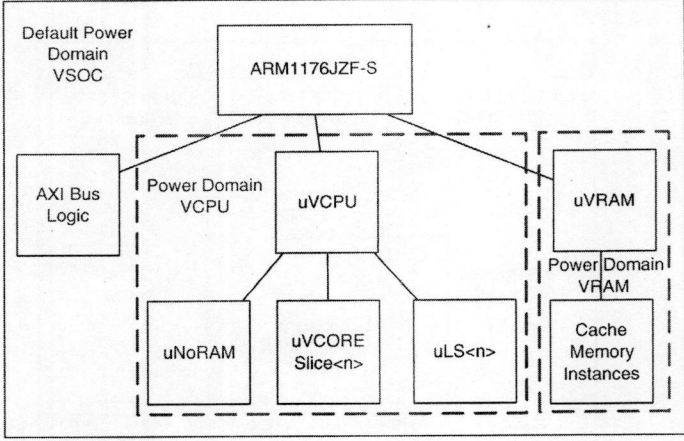

Figure 11-3   Alignment of power domain to logical hierarchy

In this example, we map the entire CPU to the VCPU domain, the whole cache – logic and RAMS – to the VRAM power domain, and the AXI bus to the VSOC power domain.

This partitioning meets several objectives:

- In some power modes, we power gate the CPU, but keep the cache powered up to enable a fast restart.
- For multi-voltage, we will operate the CPU at several different voltages, depending on work load. But in many technologies there is not enough headroom to run the cache RAMs at these lower voltages.

By maintaining a high correlation between the power domain structure and the logical hierarchy we can minimize the number of signals that cross the power domain interfaces. This minimizes the amount of logic we need to introduce to maintain electrical integrity. This in turn minimizes the QoR impact to our design and makes physical implementation easier.

For the same reasons, we have implemented the minimum number of power domains required to meet our low power objectives. Creation of unnecessary power domains makes the implementation process more complex and can negatively impact QoR.

11.1.2 Critical Path Timing

In most designs, the critical paths are known, predictable and well understood. During design partitioning for low power, we need to minimize the impact on these critical paths and make sure we do not create any new critical paths.

In our case, we have a cached microprocessor, where the critical paths are through the interfaces between the cache system and the CPU logic. However, in a multi-voltage, power gated implementation, we have placed the CPU logic and the cache memory in separate power domains. Our already critical memory interface paths are further impacted by the addition of level shifters and/or isolation logic.

These level shifters and isolation cells will be placed in dedicated geographic areas in the floorplan (voltage areas) that impose additional hard constraints on any optimization. This additional interface logic and restriction in optimization may create critical paths in the design that are not desirable.

This is also true for off-chip interfaces. Any logic in the design that is operating at a voltage level different to that of the IO will need to be level shifted prior to the IO interface thereby adding additional delay to these off-chip paths.

When partitioning a design into multiple power domains we also need to avoid the situation where logic in always-on blocks is dependant on state set in powered down blocks.

Investing the time to architect power domains carefully can help minimize the impact on physical design and system timing.

Recommendations:

- Ensure that the power domain hierarchy aligns with logical hierarchy where possible

- Pay particular attention to the critical path timing in the design – especially around power domain interfaces

11.2 Design Flow Overview

As show in Figure 11-4, the design flow for a multi-voltage, power gating design follows that of a standard implementation with a few exceptions.

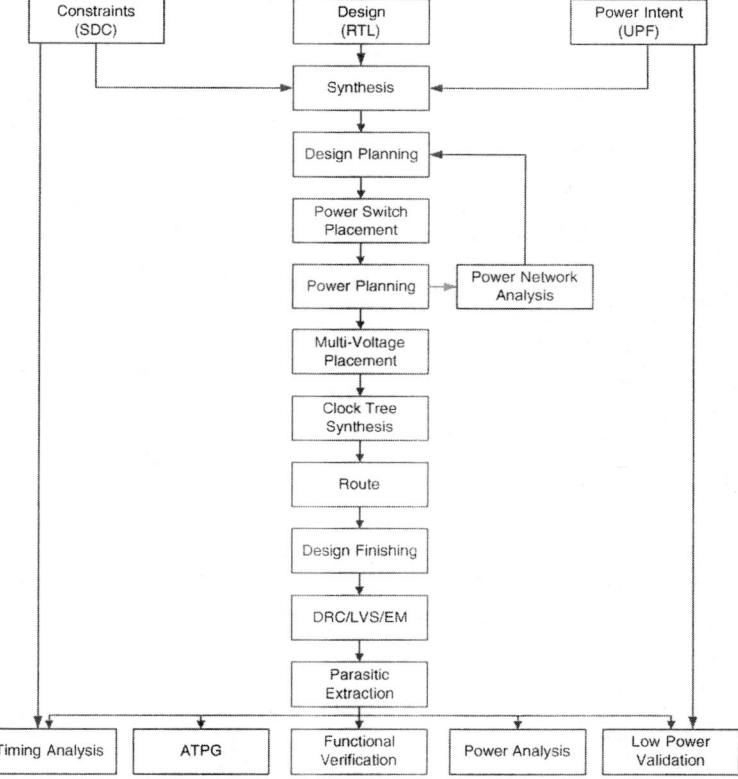

Figure 11-4   Design Flow Overview

Primarily, the differences between the standard design flow and that for a multi-voltage, power gated design are:

- Specification of the power intent
- Creation of power domains during synthesis
- State retention synthesis with always-on network management
- Multi-voltage physical design partitioning and the addition of MTCMOS switch cells
- Multi-voltage power network synthesis for MTCMOS power gating during power planning phase
- Early power network analysis to validate power gating switch topology

- Multi-voltage placement optimization including level shifter and isolation cell optimization
- Multi-voltage clock tree synthesis and optimization
- Intensive post route power network analysis with power-up sequencing verification

11.3 Synthesis

Logic synthesis is the process by which we map a generic RTL representation of the design to a target technology. In low power designs, we include the power intent – the specification of power domains, isolation, and so on – at synthesis so that optimization includes the effects of these added circuits. We can view the logic synthesis of a multi-voltage power gated design as simply a low power overlay to the typical logic synthesis process.

11.3.1 Power Intent

This power intent describes the power domains in the design and defines how power is distributed to the various power domains. It defines where level shifters, isolation cells and retention flops are required. It describes how power is switched on and off to the various power domains.

Power intent can be captured in several ways

- Through explicit definition in the RTL by hand instantiating the level shifters, isolation cells, etc.
- Through explicit definition in the RTL by using HDL pragmas.
- Through the use of tool-specific commands.
- Through the use of a accompanying power file in a standard format such as the Unified Power Format (UPF).

For the purposes of the discussion here, we will use the UPF format.

11.3.2 Defining Power Domains and Power Connectivity

In UPF, power domains are specified with the following command.

create_power_domain *domain_name*
        [-elements *list*]
        [-include_scope]
        [-scope *instance_name*]

Where:

domain_name	Identifies the new power domain
-elements *list*	Indicates which elements are in the domain.
-include_scope	Indicates that the scope of the domain should be included in the domain.
-scope *inst_name*	specifies that the scope (instance) in which the power domain is created. Default is the current scope.

As discussed previously, it is strongly recommended that the power domain hierarchy align with the logical hierarchy where possible and that the elements list be a collection of hierarchical cells within the design.

11.3.3 Isolation Cell Insertion

Isolation logic must be inserted at the interfaces of power gated blocks such that the logic in these blocks is isolated from the rest of the design during shutdown. Specific isolation cells are typically used for this purpose and these cells are available in most physical IP libraries today.

In UPF, isolation cells are specified by the following two commands:

set_isolation *isolation_name*
        -domain *domain_name*
        <-isolation_power_net *net_name* | **-isolation_ground_net** *net_name* |
        -isolation_power_net *net_name* **-isolation_ground_net** *net_name* |
        -no_isolation>
        [-elements *list*] **[-clamp_value <0 | 1 | latch | Z>]**
        [-applies_to <inputs | outputs | both>]

Where:

isolation_name	specifies the isolation strategy name.
domain_name	is the name of the domain to which the strategy is to be applied.
-elements	indicates which elements of the domain will have their interfaces isolated.
net_name	identifies the supply net(s) used to supply the isolation logic inferred by this strategy.
-no_isolation	does not isolate the port, pin, or design element specified in the elements list.
clamp_value	is the value to which the input or output shall be clamped. The default is **0**.
applies_to	indicates whether the domain's input ports, output ports, or both are isolated. The default is **outputs**.

We recommend isolating all outputs from a shutdown block. These recommendations can be met by using the default value for the **applies_to** and **-elements**.

Having defined an isolation strategy for a shutdown block, we must now also define how the isolation is controlled. The set_isolation_control command does this.

set_isolation_control *isolation_name*
      -domain *domain_name*
      -isolation_signal *signal_name*
      [-isolation_sense <high | low>]
      [-location <self | parent | sibling | fanout | automatic>]

Where:

signal_name	identifies the signal that causes the specified element to drive its clamp value.
-location	specifies where the isolation cells are placed in the logic hierarchy. The default is **automatic**.

Location determines where the level shifters will be inserted:

- Self means in the module whose output is being shifted
- Parent means in the parent of the module whose output is being shifted
- Sibling means a new module is created at the same level as the module whose output is being shifted. The shifters are put in this new module.
- Fanout means that the shifters will be place in all the destination modules of the shifted signals.
- Automatic means that the tools are free to put the shifters in the appropriate location.

There can be multiple styles of isolation cell each with specific requirements on where they are placed and how they are power routed. Thus, we recommend that the location of these cells be determined automatically by the synthesis tool based on the nature of the isolation cell being targeted. The default value of **automatic** for the **–location** switch accomplishes this.

11.3.4   Retention Register Insertion

Retention registers are synthesized automatically during the logic synthesis phase once the retention style and associated attributes have been defined.

In UPF, the state retention strategy and the specification of state retention control are separated and are specified by the following commands.

set_retention *retention_name*
      -domain *domain_name*
      <-retention_power_net *net_name* | **-retention_ground_net** *net_name* |
      -retention_power_net *net_name* **-retention_ground_net** *net_name>*
      [-elements *list]*

Where:

retention_name	specifies the retention strategy name
domain_name	specifies the domain for which this strategy is applied.
net_name	identifies the supply net(s) used to supply "always-on" power to the retention registers inferred by this strategy.
elements *list*	indicates which objects in the power domain.

Selecting a subset of elements in a design which shall have their state retained during shutdown is an architectural decision and should not be made during the implementation phase. Implementing partial state retention in a design must be a carefully managed process to ensure that the design can come out of a sleep mode correctly. The interaction and protocols between state retention functionality and the system reset controls must be well defined and honored during wakeup.

Therefore, the list of elements provided to the **set_retention** command is an integral part of the design architecture specification and should be identified prior to the implementation phase.

Having defined the retention strategy it is now necessary to define the retention control network.

set_retention_control *retention_name*
      -domain *domain_name*
      -save_signal {{*net_name* **<high | low | posedge | negedge>}}**
      -restore_signal {{*net_name* **<high | low | posedge | negedge>}}**

Where:

save_net	identifies the signal that causes the register values to be saved into the shadow registers.
restore_net	identifies the signal that causes the register values to be restored from the shadow registers.

The specific values for the options in this command are again related to the architecture of the design and the specific retention registers available in the library.

Note: The set_retention_control command also has optional arguments for defining assertions about the save and restore signals to assist in verification. See the full UPF specification of details on these arguments.

11.3.5   Level Shifter Insertion

Level shifters are inserted automatically during synthesis once the location and usage rules have been defined.

In UPF, level shifters are specified by the following command:

set_level_shifter *level_shifter_name*
     -domain *domain_name*
     [**-elements** *list*]
     [**applies_to** <**inputs** | **outputs** | **both**>]
     [**-threshold** *value*]
     [**-rule** <**low_to_high** | **high_to_low** | **both**>]
     [**-location** <**self** | **parent** | **sibling** | **fanout** | **automatic**>]
     [**-no_shift**]

Where:

level_shifter_name	is the name of the level shifter strategy, which is used by the tools for reporting.
domain_name	is the name of the domain to which the strategy is to be applied.
elements	indicates which elements of the domain will have their interfaces shifted. Default is all the interfaces.
applies_to	indicates whether to shift the (elements of the) domain inputs, outputs, or both. The default is both.
threshold	defines the voltage, in Volts, for determining when to insert level shifters. If the difference between two domains is greater than the threshold, level shifters are inserted. The default is 0V.
rule	determines whether to insert level shifters for interfaces that go from a lower voltage to a higher one, a higher one to a lower one, or both. Default is both.

We strongly recommended that level shifters be used on all power domain interfaces where an up shift is required, since this prevents crowbar currents and improves edge rates and therefore timing.

In general, we recommend that level shifters be used on all power domain interfaces where a down shift is required. It is safe to overdrive the input in the lower voltage domain. However, the timing characteristics of the destination cell will assume a correctly driven input signal operating at the same voltages the output driver of the cell. There will be an error in calculating the delay if a level shifter is not used.

Thus, we recommend **-rule both.**

The placement of the level shifters is important. The output driver for a level shifter requires more supply current than the input stage. For this reason, we recommend placing the level shifter in the destination domain – the domain that the level shifter output drives. This assures a high quality power connection to the output stage of the level shifter cell.

There are two ways to do this with the **set_level_shifter** command. We can say:

       applies_to inputs
       -location self

Or we can say:

       applies_to outputs
       -location fanout

There is no general, ironclad rule to prescribe the difference in voltage levels above which level shifters are required. This decision is technology and library dependent.

In certain cases, when the voltage difference between the two power domains is less than the threshold voltage then level shifters are not strictly required. However, when making this decision, the tolerance of the power supplies should be considered. It may be the case that when both power domains are being powered by ideal supplies that the voltage difference is tolerable, however when worst case variation between the power supplies is considered, the difference may be too great and level shifting required.

Using the default **-threshold** of 0V is a safe initial value. If timing across critical interfaces becomes a problem, we can revisit this issue and specify a different value.

In any case, we do need to specify unambiguously which power domains are at what voltage level, so the tools know where to insert level shifters. The UPF power state table constructs (**create_pst** and **add_pst_state**) provide an explicit mechanism to specify voltage levels.

In the diagram below, a net is entering the VSOC power domain which is at a different voltage from the cell driving this net. In this situation, the net must not be buffered (in the SOC domain) prior to the level shifter. Otherwise we defeat the whole purpose of using a level shifter. The requirement to avoid buffering of level shifter input effectively forces the level shifter cells to be placed close to a power domain boundary. Modern tools address these issues automatically.

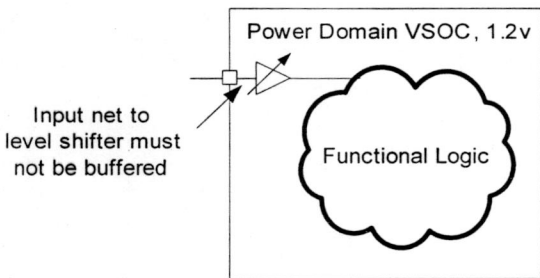

Figure 11-5   Input Nets on Level Shifters are Protected

Recommendations:

- Define a comprehensive power intent for the design that can be implemented and verified
- Define the power domains early and manage the power intent throughout the entire design process
- Define isolation logic and state retention as part of the design description (in the RTL or UPF file) to ensure full design verification
- Include isolation and state retention synthesis in the overall logic synthesis step – not as an afterthought. The area and delay cost of isolation and state retention should be an integral part of the overall cost function for the design.
- Ensure that all control networks for always-on logic (switch cells and retention registers) are buffered using always-on buffers to ensure that they remain alive during power down.

11.3.6   Scan Synthesis

As described in Chapter 5, low power designs pose special problems for scan testing.

The typical DFT challenges faced in multi-voltage design include:

- Power aware architecture and reordering of scan chains
- Automatic insertion of level shifters and isolation logic on DFT signaling across power domains
- Timing issues in scan chains that cross power domains operating at multiple voltage levels
- Intelligent routing of the scan enable signal(s) to minimize power domain crossing

Figure 11-6 shows the situation we want to avoid.

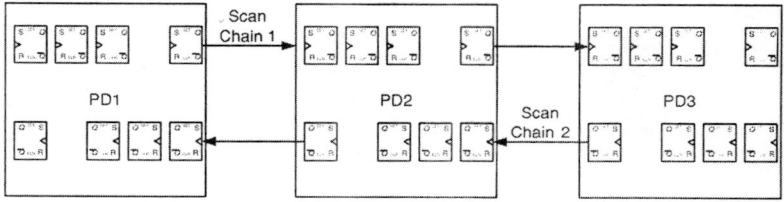

Figure 11-6   Power Domain Mixing of Scan Chains

This situation, where scan chains cross power domains, can cause real problems. At the least, level shifters and isolation cells may have to be inserted.

In the worst case, keeping all three power domains powered up during test is not possible. Since switching activity in scan mode can be much higher than in normal operation, scan testing with all domains powered up may exceed the max power for the chip.

For these reasons, we need to make every effort to implement the scan chain structure in a power aware style, where the scan paths are localized within each power domain. The scan elements can then be reordered within the power domain during placement optimization to minimize wire length. We also avoid requiring the clock skew on the scan paths to be balanced across domains, which would add unnecessary complexity in clock tree synthesis.

Figure 11-7 on page 170 shows another example of the difficulty in managing scan chains across domains. Here a scan chain comes from power domain VCPU, which is a power-down domain, and connects to Scan Flop 2. Since this connection did not exist in the original design description, the path is not isolated and therefore additional isolation logic needs to be added. This may mean that we have to add more UPF commands.

Recommendations:

- Try to implement a scan strategy that is power domain aware – minimize the number of times scan chain(s) cross power domain interfaces to restrict the number of isolation cells and/or level shifters.

- Scan chain reordering should also be power domain aware and where possible scan chains should be reordered within the power domain only.

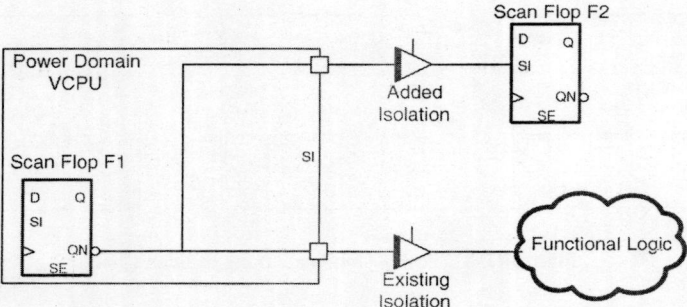

Figure 11-7   Additional Isolation Needed For Scan Chain Creation

11.3.7   Always-On Network Synthesis

When implementing any power gating design, we need to make sure that the control signals from the power gating controller remain alive during shutdown. These control signals include the controls for the power switches, isolation cells, and retention registers.

The networks of buffers that distribute these control networks are referred to as always-on networks and must use only always-on buffers and inverters. These always-on cells are very similar to their standard counterparts but are connected to the always-on supply instead of the local switched supply. This ensures that the signals on these control networks remain alive during shutdown.

In most EDA solutions available today, identification of these always-on networks is fully automated within the synthesis process. Retention registers, isolation cells and power supply switches have specific pins annotated with always-on attributes that enable the synthesis of always-on networks. Similar attributes exist on specific buffers and inverters in the library that the synthesis tool can then select to buffer these networks.

A basic power gated supply with state retention register is shown in the diagram below. The control of the power gating switch (SLEEP) and the retention register control (RETN) are both buffered by always-on networks in the design. These pins are connected to the power gating controller through a network of always-on buffers and inverters that are connected to always-on power supply VDD. The local logic in the design is connected to the switched power supply VVDD delivered by the power gating switch.

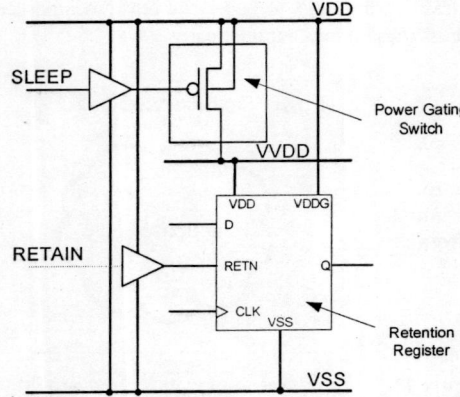

Figure 11-8   Always-on Networks Required in Power Gated Designs

Although the synthesis and physical design tools understand always-on power routing, we still need to verify the always-on nets as part of Low Power Validation at the end of the design process. The design and specification of these networks is complex; they are usually inferred from the commands that specify the isolation and retention strategies. It is easy to make a mistake in these commands but sophisticated rule-based checkers can help spot any mistakes in this area.

11.4   Multi Corner Multi Mode Optimization with Voltage Scaling Designs

It is commonplace in modern SoC designs to have a variety of functional modes. With multi-voltage design we introduce a number of additional modes to the design, including various performance levels and sleep modes, such as light sleep, hibernate and complete shutdown. These additional modes need to operate at a number of different corners (process, voltage and temperature points). The combination of these multiple modes each operating at different corners can be termed as scenarios. During the implementation process we need to both optimize and analyze the design for each of these scenarios.

For a dynamic voltage scaling design, we are concerned with ensuring full device operation at each of the various performance levels that are specified for the design. Specifically, we need to ensure that as we scale the voltage in conjunction with scaling the frequency that we have an optimal implementation at each supported combination of voltage and frequency. In a single voltage design, our worst case corner (for setup) is usually at the lowest voltage, highest temperature and worst case process. However, as we scale the performance levels in our design we may find that our worst case corner is some intermediate combination of voltage and frequency that does not reside at the edge of our operating window. Identification of this corner is important and an integral part of the implementation process.

This process is managed in the EDA tools through the use of multi-corner, multi-mode analysis. The designer can specify the modes of operation of the design and the corners at which the design is to operate. The implementation tools use these scenario specifications as constraint corners for optimization.

For the ARM1176JZF-S we intend to operate the design with varying voltage levels applied to the VCPU domain. When we vary the voltage at which VCPU operates, we also vary the clock frequency. Each of these modes, each of these combinations of voltage and frequency, has a number of associated of timing corners representing some variation in voltage, temperature and process. In addition, each of these modes will have a set of associated timing constraints, unique to each corner, which the implementation of the design must satisfy.

In using multi-corner, multi-mode techniques we need to get the best balance between quality of results and the turn time for each pass of implementation. Specifying dozens of operating modes and corners will likely result in enormous run times for the implementation tools and will not necessarily yield the best quality of results. Specifying a few well considered modes and corners for implementation followed by a more comprehensive set of modes and corners for signoff is typically the best approach.

Each defined mode needs to be accompanied with a comprehensive set of constraints (SDC) that are used during both the implementation of the design as well as the signoff.

We specify the constraints for the restricted set of scenarios at synthesis and carry them through the entire implementation flow. We then use the constraints for the full set of scenarios for signoff static timing analysis.

11.5 Design Planning

The design planning process is critical to the implementation of a low power design. It is during design planning that we are presented with the majority of challenges and have the greatest opportunity to adversely affect the QoR of the design if we are not careful. Implementing a multi-voltage power gated design is going to impact the resulting QoR. Degradation of QoR can be caused by a number of factors including:

- Reduced operating voltage for certain parts of the logic
- Added delay introduced by level shifters and isolation cells
- Increased IR drop across switch networks in the power mesh
- Increased congestion caused by the use of switch cells and retention flops
- Placement restrictions due to physically bounded voltage areas

The physical design phase of the implementation starts with a determination of the overall topology of the design. Typically, the design topology will be governed by many factors, most of which are not necessarily low power related. The location of inputs and outputs for a design will be heavily influenced by the nature of the application into which the design is to be placed. Similarly, the location of the memory system and other blocks of hard IP within the design will dominate the topology, thereby reducing the flexibility and degrees of freedom available to the designer.

We need to overlay the power intent on the physical design process such that we do not limit the resulting QoR unnecessarily or place undue restrictions on the placement optimization process.

11.5.1 Creating Voltage Areas

During design planning, we will be physically bounding our power domain with voltage areas. This will ensure that all logic that constitutes a specific power domain is placed close together, allowing power to be delivered to that logic more efficiently.

When architecting voltage areas, we need to try to optimize the relative location of functional blocks and their communication with other blocks in the design. Where possible, we want to route power consuming busses efficiently and avoid unnecessary scenic routes around power domains. As with most design planning tasks, the creation of a voltage area from a power domain will be an iterative process and will involve identifying the best size and shape of the voltage area and how best to fit that voltage area in the chip.

We determine the size and shape of the voltage area based on the performance requirements of the logic and how densely it can be packed. Also, we consider the type of logic that is to be placed - does the voltage area contain a number of hard IP blocks that would present a significant challenge to the router if placed in the middle of the design? Would using rectilinear voltage areas with these types of power domain help in the floorplanning process? We also need to make sure that voltage areas do not overlap.

Since voltage areas by definition contain logic that employ differing power strategies, care must be taken at the interface to these voltage areas to ensure that the voltage domains are truly isolated. We recommend that each voltage area be surrounded by a guard band providing some degree of physical isolation between the power domains. Guard bands present a physical placement constraint for the voltage area and they should be large enough to provide the necessary isolation but not so large that they use up valuable silicon real estate and impact the QoR. Figure 11-9 on page 174 shows how the voltage areas may be created.

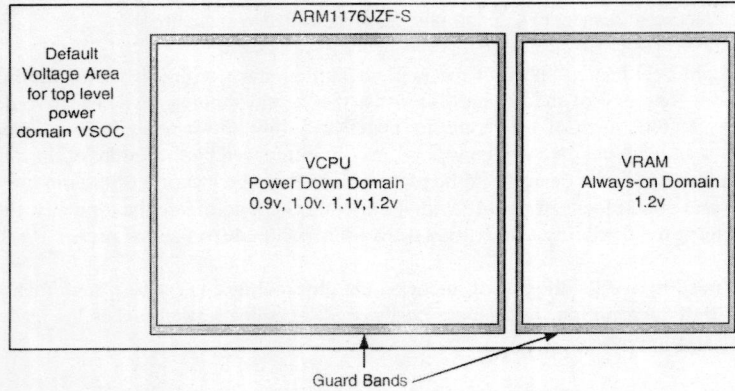

Figure 11-9   Mapping of Power Domain to Voltage Area

As mentioned in Chapter 7, nesting of power domains is a logical concept and should be avoided in the physical design. If they are absolutely required, care must be taken to ensure that the physical bounds of the nested voltage areas are separate from the voltage area of the parent power domain.

We also do not recommend splitting a power domain into multiple separate physical regions on the chip. Although this is supported in most EDA tools today, having to do this should prompt the designer to reconsider the design partitioning.

Level shifters place specific constraints on design planning. For electrical integrity, the level shifters need to sit as close to the boundary of a voltage area as is possible, as they have input signal nets that are protected from buffering. Long un-buffered nets can create timing problems in a design.

When there are many signals between two power domains, the large number of level shifters required can present a challenge. For example, power domains that provide data on a number of wide data busses will naturally require large numbers of levels shifters, all of which will be pulled as close to the physical boundary of the voltage area as is possible. This can result in multiple site rows being consumed with level shifters around the periphery of a voltage area, leading to degradation in QoR for those nets that have their level shifter placed furthest from the boundary.

In these situations, it is recommended that power domains be floorplanned in such a way as to align data busses across power domain boundaries and give as large a dimension as possible to common interfaces. This will allow for spreading of the level shifters, resulting in a smaller physical separation between the logic in each power domain.

11.5.2   Power Gating Topologies

Where local power gating is employed, there are a number of alternative topologies for the placement of the power gating switches. These are discussed in detail in Chapter 14.

For most designs, including the ARM1176JZF-S, we use a grid design for the power network. With modern tools, we have two options for specifying the placement of the switches. We can specify an x and y increment, and the tool will place the switches evenly throughout the network. Or we can give the tool a design goal (such as current capability or IR drop) and the tool can do switch placement exploration and return with a list of options for the designer.

The size and placement of the switches is critical in determining the IR drop across the switch network. This in turn is critical in determining timing, since delay is very power-supply dependent.

In UPF, power gating switch cells are specified by the following command

```
create_power_switch switch_name
        -domain domain_name
        -output_supply_port {port_name supply_net_name}
        {-input_supply_port {port_name supply_net_name}}*
        {-control_port {port_name net_name}}*
        {-on_state {state_name input_supply_port {boolean_function}}}*
        [-on_partial_state {state_name input_supply_port {boolean_function}}]*
        [-ack_port {port_name net_name [{boolean_function}]}]*
        [-ack_delay {port_name delay}]*
        [-off_state {state_name {boolean_function}}]*
        [-error_state {state_name {boolean_function}}]*
```

Where:

switch_name	specifies the name of the switch instance to create;
domain_name	specifies domain containing the switch.
-output_supply_port	identifies the output supply port of the switch and the net where this port connects.
-input_supply_port	identifies the input supply port of the switch and the net where this port connects.
-control_port	identifies a control port on the switch and the net where this control port connects.
-ack_port	identifies the acknowledge port on the switch and the signal net where this port connects.
-ack_delay	identifies the acknowledge port and delay on the switch where this port connects.
-on_state	defines what value of the control port turns the switch on.
-on_partial_state	defines what value of the control port turns puts the switch in a current-limited state.
-off_state	defines what value of the control port turns the switch off..
-error_state	identifies any error states, which if defined on the switch can be flagged during simulation or analysis.

11.5.3   In-rush Current Management

The power-up sequence for a power gated design is critical for limiting the in-rush current, so that we do not cause voltage spikes that could corrupt retention registers or data in adjacent domains.

One effective technique, employed on ARM1176JZF-S, uses a buffered switch which provides a buffered (SLEEPOUT) version of the switch control input pin (SLEEP). This allows daisy chaining of the sleep control network throughout the design by connecting the SLEEP pin of one switch to the SLEEPOUT pin of another switch.

The exact topology of this daisy chain can be user defined, but the end goal is the same – a buffered sleep network that can be turned on quickly but at the same time limiting the in-rush current. This usually entails grouping sets of switches together, and driving the whole group with the same SLEEP pin, so they all turn on at the same time. One of the SLEEPOUT signals from the group is connected to the SLEEP pin of the next group. Thus, groups of pins are daisy chained. The speed vs. noise trade-off now involves determining the size of the group. This can be done through power analysis.

11.5.4   Recommendations:

- Voltage areas should be guard banded where necessary.
- Placement optimization must be constrained to ensure that cells in a power domain are placed inside the associated voltage area.

- New cells created during the physical design optimization process must be placed in the appropriate voltage area and powered by the correct power supply.
- Short feed-through paths in a voltage area should be protected from repeater insertion during design optimization and design-rule fixing to maintain wire connection.
- The isolation cells and level shifters should be placed at the voltage area boundary to ensure minimal impact to timing.
- It may be necessary to experiment with various power switch topologies and compare the results to determine which is most suitable for the design in question.
- Consider the power routing impacts for the switch topology chosen.

11.6   Power Planning

With a physical topology in place we can now begin to consider how to provide power to the various voltage areas in the design. Since by definition, each power domain employs a different power strategy, it is likely that we will be routing different power rails to each voltage area.

Minimizing the voltage drop across each of these power rails is a key part of meeting the performance goal. Unfortunately, many of the techniques that we employ in a low power design can make the voltage drop and noise problem worse. For example, when the design is operating at reduced voltage levels, our available margin for voltage drop is considerably reduced. When we change performance levels or power-up a block after shutdown, the on-chip current gradients become worse which can lead to noise injection. And in power gating blocks, there is an IR drop across the power switches.

Most EDA tools provide some level of automated power network synthesis (PNS) capability for the distribution of power across a design. We recommend highly that the implementation of a multi-voltage design utilize this type of PNS capability wherever possible. The number of concurrently changeable variables in a multi-voltage system is such that attempting to construct a power plan manually may result in a sub-optimal result both in terms of routing resources consumed and the voltage drop on the rails.

Power network synthesis allows the designer to specify the absolute constraints on the power plan (such as maximum voltage drop, routing layers and via requirements). It ensures that the power budget for the design is satisfied and provides integrated voltage drop and electro-migration analysis. Most level shifters available today that are shifting from a low voltage to a higher voltage require two power supplies. Cells that combine level shifting and isolation require a secondary supply that provides always-on power to the cell. During the power planning process it is necessary to hookup the primary power pins to the appropriate voltage rails. There are various ways that this can be achieved depending on which EDA tools are being used. For example, for isolating level shifters, the primary power supply pins can be hooked up to the appropriate voltage rails through PNS and the secondary power pin (for always-on power) can be connected by the power router later in the design flow when the final placement of these cells is frozen.

The sleep transistor power network is composed of three components: a permanent power network, a virtual power network and an array of the sleep transistors. All of these components contribute to the quality of the sleep transistor power network in terms of IR drop, routing resources and silicon area. Consequently, the synthesis of the sleep transistor power network becomes a challenge. Historically, most power-gating designs used scripts to insert and place the switch cells, and create power straps and rails based on heuristic rules. Today, the sleep transistor power network can be synthesized using a power planning tool PNS [PNS] that considers all the three components.

Recommendations:

- Use a power planning tool which supports switch cells to create the sleep transistor power network including switch cell insertion and placement.

- Run both static and dynamic IR drop analysis on a design with switch cells to verify the power integrity. Significant static IR drop violations usually require increasing the number of switch cells and adjusting their positions.

- Dynamic IR drop violations are commonly fixed by decoupling capacitor insertion. Accurate power-on sequence I-V responses have to be obtained by transient analysis to assure that all IR drop violations are fixed.

- In the case of any IR drop violations, try to fix the violations in the permanent power network, because any changes in the switch cells distribution and virtual power network usually require re-analyzing wakeup latency and peak current.

- Run wakeup in-rush current analysis to check that the max in-rush current and wakeup latency meet the defined constraints. For small designs, it is possible to run this analysis in SPICE. At chip level, the wakeup analysis is done using rail analysis tools.

- Run dynamic IR drop analysis covering the wakeup period to detect any violations caused by current spikes, through common power rails, on the parts of the design that are alive while the power-gated block is waking up.

11.6.1 Decoupling Capacitor Insertion

An integral part of the power planning process involves ensuring that the noise coupling between different power supply rails and between cells in the same power domain is minimized. This power and ground noise can degrade timing and possibly cause functional failures. Inserting decoupling capacitors (decap cells) in the design is the method commonly used to address these issues.

Decoupling capacitor insertion becomes more challenging in power-gating designs than in normal designs. Implementing in-rush current control techniques, such as daisy chaining the power-gating structure to sequentially charge the power-gated block, reduces the in-rush current. However, the turn-on current of each switch, particularly of those switches which are close to the head of the daisy chain and turned on first, can still generate significant power and ground noise pulses in the rest of the design, through the permanent power supply and common ground networks. Adding decoupling capacitors to the permanent power supply nets is an effective way to reduce this noise. It is most effective to insert the decoupling capacitors next to the switch cells which are the sources of the noise during wakeup.

Apart from the power and ground noise generated by the in-rush current, there is also the noise generated by cell switching currents in the power-gated block during normal operation. Decap insertion is a common method to address this noise. However, decap insertion becomes more challenging in the power-gated design. In sub-90nm technology, decap cells exhibit high leakage power that needs to be controlled. Also, the decap cells added to the virtual power network have significant impact on the wakeup latency and in-rush current. The larger the decoupling capacitance, the larger the in-rush charge current and the longer the charge time (latency) at wakeup. Therefore, the optimization of de-cap insertion in the power-gating design becomes very important to achieve maximum noise reduction with minimum added capacitance at the virtual power network. This can be done by identifying noise hot spots using dynamic IR drop analysis tools and then inserting just enough capacitance at the hot spots to reduce the noises meeting defined noise target. It is worth noting that advanced EDA tools have implemented such optimal decap insertion methods.

Recommendations:

- Add as much decoupling capacitance as permitted in the permanent power network at the positions close to the switch cells. This achieves the maximum effectiveness and minimum impact on the wakeup latency and in-rush current. It is convenient to integrate the decap into the switch cell to simplify decap insertion. The maximum capacitance of the decap on the permanent power network is constrained by the leakage and area penalties.

- To fix dynamic IR drop violations in the post-layout stage, it is preferable to add decoupling capacitance to the permanent power network close to the violation spots, if the violations are related to the permanent power network. The rest of the violations have to be fixed by adding decap to the virtual power network at the violation sports.

11.7 Clock Tree Synthesis

The impact on the overall power consumption of the clock tree in a design is significant. In many cases, more than half of the overall power consumption of a design may be due to the clock network. There are a number of ways to mitigate the effects of the clock tree power, including clock gating and minimizing clock tree insertion delays.

However, when we are dealing with a multi-voltage design, we have additional restrictions on how we can manipulate the clock tree to meet both our low power requirements and our performance requirements at each performance level. In a single clock multi-voltage design, the clock is used in multiple power domains and hence crosses a number of voltage area boundaries. As the clock passes through each voltage area its latency is modified based on the voltage at which those voltage areas are operating. If the clock path buffering and data path buffering are not well balanced across voltage areas then skew management becomes very difficult.

The diagram in Figure 11-10 illustrates the issue. In this particular situation, both the data and the clock are being sent by the CPU to the memory. The data path buffering is split between the VCPU and VRAM power domains with the clock path buffering residing primarily in the VRAM power domain. Assume that we have very good skew minimization when both power domain VCPU and VRAM are operating at the same supply voltage. When the voltage on power domain VCPU is reduced as part of performance scaling then we run the risk of slowing the data path relative to the clock and introducing setup violations in power domain VRAM.

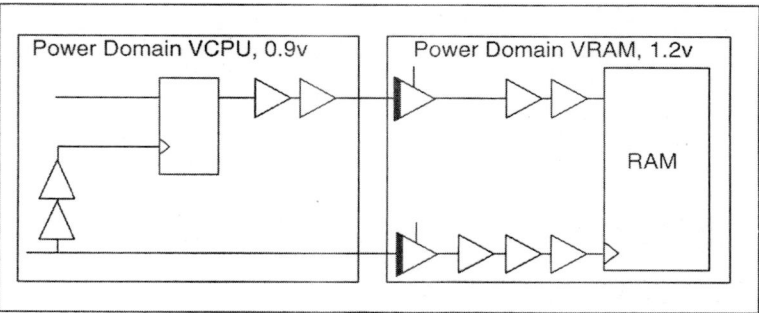

Figure 11-10   Clock Tree Synthesis Challenge in a Multi-Voltage Design

In order to help alleviate this problem, clock tree synthesis algorithms in current EDA tools are multi-voltage aware and use a bottom-up approach to constructing the clock tree. Each voltage area is processed in turn and the clock networks for each voltage area are constructed to minimize the skew. These low level clock tree structures are then merged by the construction of higher level clock trees that join together the subtress to form an overall clock network for the design.

Adopting this approach helps locate the buffering for the clock tree in the same voltage area as the data path buffering. This clock buffer clustering prevents the clock network from winding its way in and out of a number of voltage areas, requiring level shifters to be inserted every time the clock network crosses a voltage area boundary. This level shifter insertion process would add significant delay to the clock tree and impact the ability to minimize the clock skew and power. Figure 11-11 illustrates this clustering approach. Note that the VRAM and VPCU domains have level shifters ("LS" in the figure) in their clock trees.

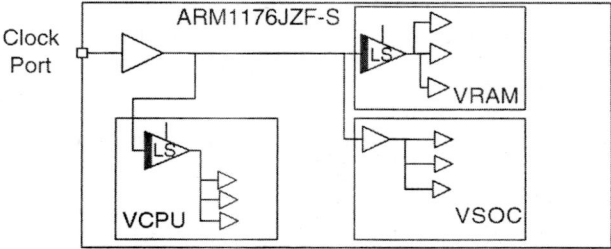

Figure 11-11   Bottom Up Clock Tree Clustering

This approach lends itself well to voltage areas that have a good mix of clock and data buffering such as blocks of standard logic. However, when we have a voltage area which contains little standard cell logic but rather many blocks of hard IP, extra care must be taken to ensure that the imbalance of clock tree and data path buffering will not introduce significant skew problems as voltage levels are scaled.

Routing

The leading commercial routers today are all low-power aware and will honor the power intent in the design. There are however some potential pitfalls that can occur during detailed routing which can be avoided by intelligent floorplanning and design partitioning.

When a design is partitioned into multiple voltage areas, hard placement and routing restrictions exist that can impact the routing of the design. This can lead to degradation of QoR. It is therefore important to consider routing of the design, especially on-chip bus structures and global control networks, when floorplanning the design.

If the best route for a net from one part of the design to another is across one or many voltage areas then a decision has to be made on how to deal with the net.

- Route the net through the voltage area(s) and then add the appropriate level shifters to each voltage area crossing
- Detour the route around the voltage area(s) to the destination

With the first option, adding level shifters to every point at which the net crosses a voltage area will be expensive in terms of area, power and delay. The second option will introduce an incremental delay compared to a direct connection that would be available in the single voltage design; however the impact of this approach is considerably less than with the first option.

Both of these options will impact the QoR and so clearly we want to avoid this situation. The cost (in time) of rectifying this situation at the detailed routing stage will be expensive and so avoiding this situation by considered design planning earlier in the design cycle is recommended. Complete avoidance of this situation in a multi-voltage design may not always be possible but making every attempt to minimize the problem is recommended.

Figure 11-12 on page 183 illustrates the detour routing problem. Here a net going from power domain A to power domain C must cross power domain B which is at a higher voltage. Rather than adding level shifters to this net and routing it through power domain C, we can detour route this net around power domain C.

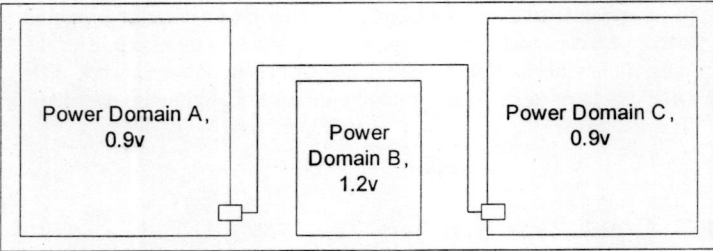

Figure 11-12   Detour Routing Around a Power Domain

11.8 Power Analysis

Having completed the implementation of our multi-voltage, power gated design, it is now necessary to verify the integrity of the power network. Two aspects of our power network are critical: the voltage drop seen by the standard cells in the power gated blocks and the profile of the in-rush current during power-up sequencing.

Performing power rail analysis of the design is critical. There will be many power rails in the design and the integrity of each needs verifying and the interaction between rails (coupling) also needs to be verified. Using a comprehensive rail analysis tool, we can perform extensive analysis of the power networks in the design. The coarse level power network analysis was completed during the power planning phase. What we are looking for here are areas in the design where we have excessive voltage drop due to local level clustering which results in power "hot" spots.

Implementation tools today are very good at performing local clustering of registers, clock gates and clock tree buffers to ensure that aggressive timing goals are met. This however can make the voltage drop problem worse in those regions of the chip, especially in regions with switched power rails. Since logic spreading in these areas is not viable (this will impact the QoR) we must deal with the voltage drop problem in a different manner and typically sizing the switch cells and power mesh is the best approach.

The rail analysis gives us valuable data on the voltage drop seen by every standard cell in the design. Where we have an unacceptably large voltage drop we can try sizing or duplicating power switch cells and sizing the power mesh to alleviate the problem. Creation of instance specific voltage drop data for a design can then be used by a signoff static timing to measure the impact of voltage drop on the timing of the design.

The second area of power analysis that needs to be addressed is that of in-rush current. We have taken great effort in architecting a power gated design that supports a fast transition from functional mode to standby mode, and maintains the necessary state to resume functional mode quickly when the design comes out of sleep. Recovering from a sleep mode quickly is important, but not so quick as to induce a large in-rush current. By performing extensive transient analysis we can model the power-up sequencing of the design under various conditions and determine the optimal balance between wake-up speed and rush current management.

11.9   Timing Analysis

A multi-voltage power gated design will need to function in a number of different modes each operating at different corners. During implementation, we selected a reduced set of scenarios - operating modes and corners – and used these as constraints. Hopefully we chose the best and worst cases and produced a design that will work under all scenarios.

The purpose of signoff static timing analysis is to prove that we have, in fact, implemented a design that works under all scenarios. As such it is necessary to accurately specify all of these scenarios and provide all the necessary technology libraries required to analyze the design at the supported performance levels.

In particular, we need to make sure that the libraries – especially level shifters – are characterized for the voltages as which we run signoff static timing analysis.

Run times will now be significantly longer than for a design that is not implemented with aggressive low power techniques.

During the fabrication process for a CMOS based design, it is expected that there will be minor variations in resistance and capacitance of both the transistors and the metallization. These variations will impact the timing paths in different regions of the chip. We recommend that a small window of uncertainty be placed around each of these operating corners to provide a greater level of confidence in the final signoff results. This window of uncertainty can help counter any on-chip variation (OCV) that may exist across the die. On-chip variation will derate the path delays in unfavorable directions, such as speeding up clock and slowing down data for setup checks and visa-versa for hold checks.

11.10 Low Power Validation

Having successfully implemented our multi-voltage power gated design that meets all power and performance targets, it is now necessary to validate the integrity of the design. Specifically we want to ensure that the low power intent provided at the start of the implementation process has been successfully implemented in the final design.

We validate functionality and the integrity of our low power implementation using three methods:

- Gate level logic simulation
- Equivalence checking
- Rule-based methods

Clearly gate level simulation of our final design can tell us if the design still remains functional with the low power structures. Specifically, we can validate that the design:

- resets cleanly at startup
- can be placed into various sleep modes
- behaves appropriately during shutdown
- powers-up successfully after shutdown

Formal equivalence checking tools can prove that the gate level netlist is equivalent to the original RTL plus UPF code.

A rule-base tool can tell us that the power structure of the final gate level netlist makes sense. It can validate that the isolation cells and level shifters are placed in the correct domains, and that all nets requiring isolation or level shifting have the appropriate cells in place. It can check that cells that required always on power – such as retention registers, isolation cells, and buffers of power control signals – do, in fact, have the appropriate supplies. Finally, we can use these tools to find situations where there are redundant isolation cells or level shifters.

11.11 Manufacturing Test

In a typical voltage scaled system, various parts of the design will be running at reduced power supply levels. During the implementation process, the design can be optimized at these voltage levels thereby yielding an overall lower power design that meets the needs of the system application. However, in most manufacturing test situations, the design will be tested at nominal supply rail voltage levels which will cause the design to consume significantly more power during test than it would during normal operation. If the design has only been packaged assuming functional power dissipation numbers then thermal failure of both the design and package at test can result. In addition, this increased power consumption in the core can lead to significant voltage drop in the power distribution network which in turn leads to functional failure. It is therefore necessary to consider the requirements of both the functional and test modes of the design when implementing voltage scaling.

When performing manufacturing test on designs implemented in very small geometries (90/65/45 nm) it is necessary to use a variety of test models and test strategies. Specifically, testing the design with stuck-at faults models now needs to be complimented with transition fault delay testing. Delay fault testing is particularly susceptible to changes in the voltage and frequency at which a design operates. Similarly, path based testing is also dependent upon the voltage at which the design is being operated. As will be discussed later in this chapter, the critical path of a voltage scaled design is not the same path at each voltage level at which the design operates. Each performance level supported by the design may well have a unique critical path. Each of these paths needs to be tested during at-speed test in a manufacturing test environment.

However, as we reduce the supply voltages at which we test our design, delay behavior changes dramatically. In particular, weak circuit elements and defective circuit elements do not behave in the same way. In addition, the tester environment differs from the system environment in noise content, heat dissipation and pin load. As a result, care needs to be taken to distinguish between correct behavior and defective behavior. In many cases, final decisions on correct timing and voltage levels will need to wait until silicon is available.

Automating the manufacturing test process will require performing stuck-at fault testing and delay fault testing at several operating voltages, and also may involve both nominal and high temperature test. Developing the appropriate mix of tests in order to guarantee sufficient quality at reasonable cost is a complex problem. Different trade-offs between test time and coverage will apply to different products.

CHAPTER 12   *Physical Libraries*

One of the first steps in implementing a low power design is to select a library of standard cells and a set of memory compilers that support the low power strategy used in the design.

This chapter describes the requirements for standard cell libraries and memories for a multi-voltage, power gated design.

12.1 Standard Cell Libraries

Standard cell libraries are tuned for different performance, power and area goals. For low-power design the choice and mix of libraries may have a significant impact on power, timing and area.

One key characteristic of a cell library is cell height. Cell height is measured in tracks, which is the metal one (M1) pitch. An 8-track cell is tall enough that eight horizontal M1 wires can run through it.

Cell libraries are designed to a certain number of tracks in height, and this height affects the timing and routing characteristics of the library:

• Tall track height libraries support more complex routing, larger drive strength transistors and typically are tuned for performance – but may exhibit higher leakage power. An 11 or 12-track library is considered a tall track height library.

- Low-track height libraries are optimized for area efficiency, but generally are designed with smaller, lower drive strength transistors so are less appropriate for high-speed designs. A 7 or 8-track library is considered a low track height library.
- Standard track height libraries are designed to give a reasonable trade-off between area efficiency and performance. These libraries are used in the majority of designs. A 9or 10-track library is considered a standard track height library.

Libraries can be built with compatible footprints using transistors with different threshold voltages:

- High-V_T libraries exhibit the lowest leakage power at the cost of somewhat lower performance. High-V_T libraries are a good choice for non-timing-critical designs, and for non-critical paths in higher performance designs.
- Low- V_T libraries are built with high-speed but leaky transistors and are tuned for high performance. They dissipate higher static and dynamic power as a result.
- Regular- or Standard- V_T libraries sit in between these and offer lower performance than the Low-V_T transistor versions at reduced leakage and dynamic power.

Every threshold variant adds mask costs for different implant layers and introduces some extra variability into the final silicon. Typically the designer would limit usage to, say, two threshold variants only.

Libraries can also be further optimized for low static power

- Long channel-length gates can be used to reduce leakage, at some cost in terms of timing and area.
- The "stack-effect" of series transistors inside gates can be exploited to reduce source-drain leakage across the other transistors for more complex gate structures.

12.1.1   Modeling of Standard Cell Libraries

Library-level IP abstracts the detailed characteristics of the underlying circuits into cell-level models to allow implementation and verification without the transistor level models. This level of abstraction provides "front-end" library design views where the commercially sensitive internals of the cells or memories are removed. A set of views that convey the port-level interface, the functionality, timing, and power cell is sufficient for synthesis, place and route, parasitic extraction and post-layout timing analysis. For manufacture the "back-end" library cell views are switched in with the technology dependent and commercially sensitive transistor-level layout.

The abstract views required include:

- Timing models – to support multiple corner synthesis, optimization and analysis.

- Physical models – in the form of layout "abstracts" with all power and signal ports
- Functional models – for gate level netlist simulation
- Power models – to support dynamic and leakage power optimization and analysis
- Test models – to support ATPG and ensure fault coverage

Multi-voltage power gated designs rely on standard cell libraries to provide the modelling information the tools need to optimize and analyze timing and power.

12.1.2   Characterization of Standard Cell Libraries

Historically, standard cells were characterized at a number of process, voltage, and temperature conditions. Several copies of the timing models were shipped: for example, a worst case (slow process, low voltage, high temperature), best case (fast process, high voltage, low temperature) and typical. Worst case timing was used for checking setup times and best case for hold times.

With the move to 90nm technology (and below) and the adoption of aggressive power management techniques, characterization has become much more challenging.

At 90nm and below, wires are becoming more resistive, to the point where network impedance can be higher than the output impedance of the driving gate.

Temperature Inversion

In older processes, gate delay always increases with increases in temperature. But starting at 90nm, it is observed that the gate delay decreases with an increase in temperature under low VDD or slow signal transitions. This is known as temperature inversion.

The physical nature of the temperature inversion is complex, but here is a general explanation of the phenomenon based on device behavior.

Gate delay is directly correlated with saturation current (I_{DSAT}); the larger I_{DSAT}, the smaller the gate delay. I_{DSAT} increases linearly with carrier mobility and quadratically with voltage headroom (VDD – V_T). As temperature increases, mobility decreases while the voltage headroom increases since V_T drops at higher temperature. The temperature effect on the gate delay from the change in mobility is opposite to that from the change in voltage headroom. The overall effect determines the gate delay.

In older processes, the effect of temperature on mobility dominates, due to a large voltage headroom. Consequently, gate delay increases with temperature with a given input transition and output load on the gate.

In sub-90nm nodes, VDD is scaled significantly lower than in the older processes. As a result, the voltage headroom becomes small enough that the gate delay becomes more sensitive to the change in V_T than the mobility.

At high temperature, the delay reduction due to the decrease in V_T overwhelms the delay increase caused by the decrease in mobility. This results in a smaller delay at high temperature and hence temperature inversion. The lower the VDD, the higher delay sensitivity to V_T and hence the stronger the temperature inversion.

Temperature inversion also depends on signal transitions. The slower the transition, the longer the period of the gate transition and hence the stronger the temperature inversion behavior.

In any given library, the effect of temperature inversion is not consistent from cell to cell. It also varies from timing arc to timing arc in a particular cell. Thus, each cell and every arc from every input to the every output must be characterized fully to include the effects of temperature inversion.

New Library Timing Models

With multi-voltage, voltage scaling, and power gating designs, the supply voltage may vary significantly from gate to gate or module to module. With traditional library modeling techniques, interpolating delay values from given timing/voltage data can be quite inaccurate.

For all the above reasons, new library models are needed. The traditional approach of modeling gates as time-dependent voltage sources with a series resistor is too inaccurate. With composite current source modeling (CCS), the gate output is modeled as a time and voltage dependent current source with essentially an infinite drive resistance. This approach provides accurate timing estimates for a wide variety of load impedances. It is accurate enough that instance-specific delays can be calculated for a specific voltage – addressing the problems of multi-voltage power gating designs.

Current source modeling also can model temperature inversion, addressing (but not solving) one of the big problems for sub-100nm designs. Finding the worst case temperature is still a difficult problem.

12.2 Special Cells - Isolation Cells

Isolation logic is needed at the interface between the powered down and powered up domains. Isolation ensures that there are no floating inputs to the active power

domains, which could result in crowbar currents. It also assures that the inputs are in appropriate logic states.

The isolation logic can be implemented either in the powered down domain to control output signals (output isolation) or in the active power domain to control input signals (input isolation). This section describes isolation circuit implementation guidelines.

12.2.1 Signal Isolation

Three types of isolation circuits can be inserted at the outputs of a power-down island:

- clamp the signal to "0"
- clamp the signal to "1"
- clamp the signal to the last value.

For an output that requires clamping the signal to "0", we can use a NAND gate and an inverter, as shown on the left in Figure 12-1, for signal isolation. This design uses an active-low isolation control signal which forces the output low even if the other input floats. The circuit diagram on the right in Figure 12-1 shows why the circuit is immune to a floating signal on IN. As long as ISOLN is low, the bottom transistor is off, no current can flow through the gate, and the input to the inverter is pulled up.

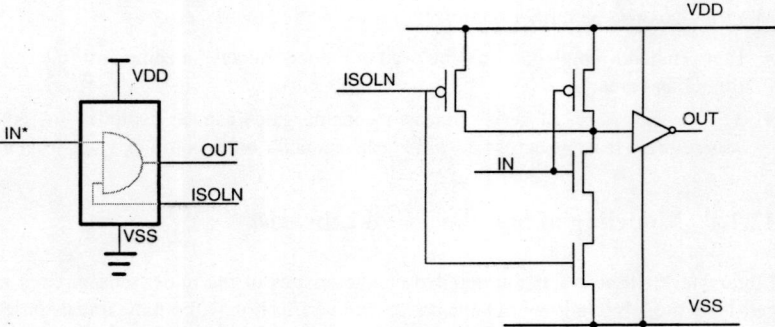

Figure 12-1   AND-style Isolation Cell

For an input that requires holding a logic "1" when the source power domain is powered down, we can use a NOR gate, as shown on the left in Figure 12-2, for the signal isolation. This is shown with an active-high isolation control signal which forces the output high even if the other input floats. The circuit diagram on the right in Figure 12-2 shows why the circuit is immune to a floating signal on IN.

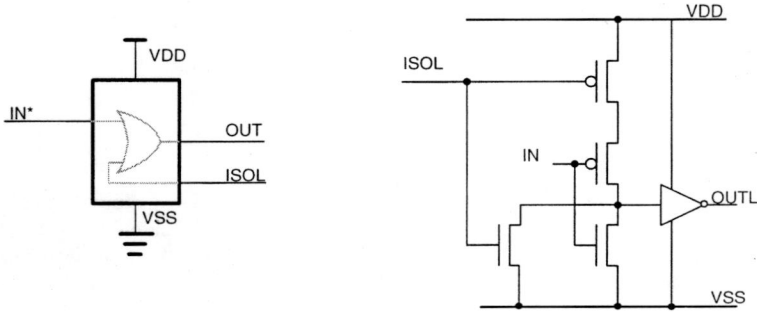

Figure 12-2   OR-style Isolation Cell

Figure 12-3 shows one example of the third type of isolation cell, which includes a retention latch to retain the state of the output signal.

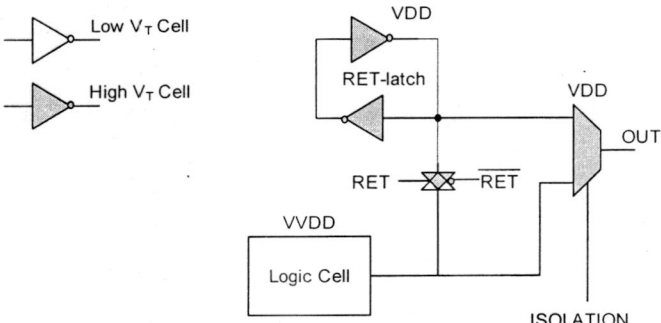

Figure 12-3   State Retention Isolation

The retention latch is controlled by a pulse signal RET which is asserted just before the logic cell goes into sleep to save the current output state into the retention latch. Then the isolation control ISOLATION is asserted to switch the output mux to the retention latch and the logic cell goes into sleep where the virtual power VVDD is shut off.

The isolation control signal is usually distributed as a global signal across power domains. To ensure the signal is alive when one or more domains are powered down, the isolation control signal is distributed with an always-on buffer tree.

12.2.2   Output Isolation vs. Input Isolation

Output isolation has some significant advantages over input isolation.

For an output signal that goes to multiple different power domains, only one isolation cell is required with output isolation. With input isolation, each destination would require its own isolation cell.

With output isolation, all the isolation cells in a domain share a common control signal. With input isolation, a block may require multiple isolation control signals – one from each domain from which it gets an isolated signal.

Output isolation has one drawback, which is that custom isolation cells are required. Although they function as AND or OR gates, isolation cells require always-on power. Most standard cells connect power and ground through abutment. But in a power gated domain, one of the supplies connected by abutment is switched. Therefore (output) isolation cells require special physical design to accommodate connection to the always on supply.

12.2.3   Sneak DC Leakage Paths

One potential problem for isolation circuits is the potential for sneak path leakage. An isolation cell clamps the output of a powered-down domain at either a "0" or "1" state by the pull-down or pull-up transistor respectively. The pull-down/up transistor provides a possible DC path to VDD or VSS through the powered-down outputs from logic in a connected alive power domain. Such a situation is illustrated by the diagram in Figure 12-4.

In this case, the powered-down output is clamped at "0" and drives one of the inputs of an XOR gate in an alive power domain. When another input of the XOR gate is "1", a leaky DC path is formed from VDD to VSS through conducting PMOS transistors and leaky transmission gate in the alive domain and the pull-down NMOS in the powered-down domain. Although it is not a conducting DC path, the high leakage of the transmission gate results in considerable leakage current in the sneak path. This can defeat the purpose of power-gating. It is worth noting that cells in a power-gating domain are commonly low-V_T cells for high performance at the expense of high leakage. Therefore, the OFF state transistors in any DC path could cause considerable leakage.

Commercial libraries do not have transmission gates on cell inputs to avoid leakage paths as well as for a variety of other reasons.

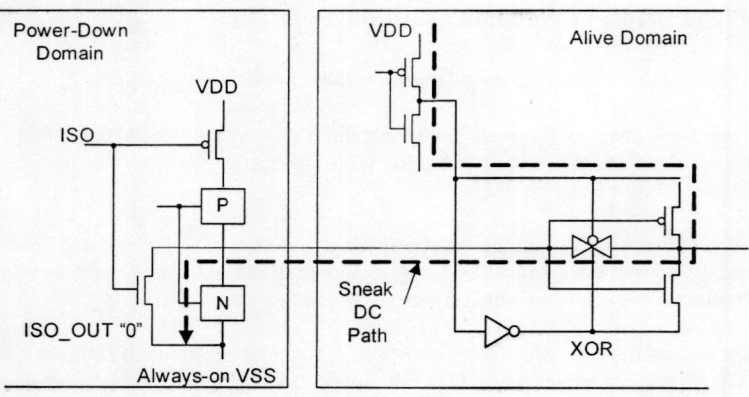

Figure 12-4   Sneak DC path through leaky XOR gate

12.2.4 Recommendations

- Output signal isolation method is usually a better choice than input isolation method due to fewer isolation cells and simpler isolation control.

- If custom output isolation cells are not available, it is feasible to use the standard cells (AND, NOR). However, an always-on power area must be created next to the power-down island, and the standard cell based isolation cells must be placed in the always-on area. This allows the cells to get power when the powered-down island is in sleep mode.

- Isolation cells introduce delay penalty. Therefore, they should only be inserted where necessary. We recommend analyzing the power-down relationship between power domains to identify those power domains that go to sleep and wake up together. Interfaces between such power domains need not be protected by isolation cells because there is no circumstance in which one side of the interface is powered-down and another side is alive.

- During placement and physical synthesis, it is important to ensure that the input and output isolation cells are placed inside the power island and close to the power island boundary in layout. Moreover, the interface nets must be protected so that no logic cell can be inserted in the nets. Such insertion between the isolation cells and power-island's ports will defeat the purpose of the isolation cells which needs to be connected directly to the ports to isolate the interface signals.

- Pass-gate logic cells should not be implemented at the interface between two power domains. This constraint is necessary to prevent sneak DC paths from VDD to ground through the pass-gates and the transistors in the interface logic cells.

- We recommend checking the logic cells at power-domain inputs to make sure the defined isolation states do not cause any sneak DC leak paths through the interface cells.

12.3 Special Cells - Level Shifters

When signals cross voltage domain boundaries and logic level switching voltages are not the same, level shifter cells must be inserted to convert the signal voltage to the correct voltage at the receiving domain. There are two cases. Shifting the voltage down is simpler; shifting up is more challenging and adds extra complexity.

When shifting down, we make the assumption that the higher voltage (VDDH) is not higher than 25% above the nominal voltage of the cells used in the lower voltage (VDDL) domain. Excessive voltage can accelerate time to failure; keeping within 25% of nominal is a reasonably safe overdrive level.

In the case of shifting down the level shifter cell can then be just a simple inverter or buffer. CMOS gate inputs can be driven higher than the power supply voltage without problems, up to the gate breakdown voltage. We power the level shifter from VDDL and drive the input from 0 to VDDH; the output swing will be 0 to VDDL. However it is important to properly characterize the level shifter with the actual voltages that will be present on each pin, in order to account for the transition times on the high voltage input.

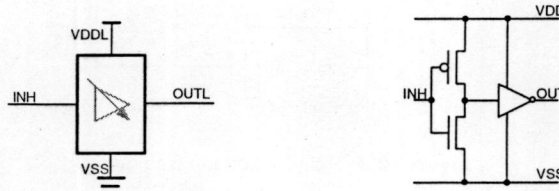

Figure 12-5   High to Low Level Shifter

In the case of shifting up it is necessary to design a special level shifter circuit because a low voltage swing input signal would not necessarily be strong enough to turn the NMOS input transistor fully on. This could lead to an unacceptably long risetime or falltime. A simple "low to high" level shifter that solves this problem is shown below. The input and the inversion of the input drive a simple amplifier.

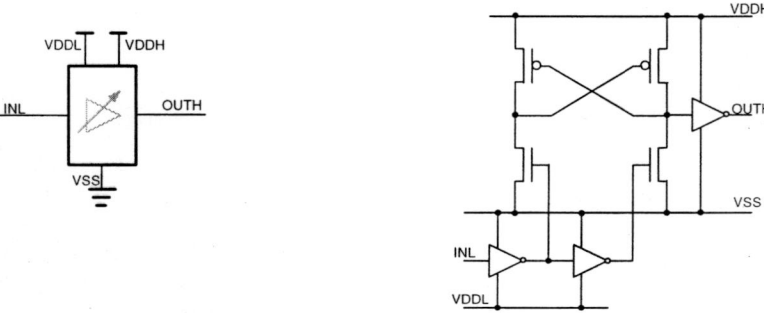

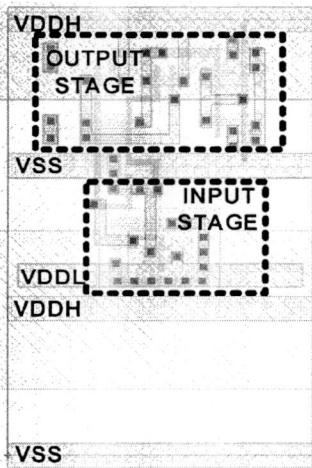

Figure 12-6   Low to High Level Shifter

Due to voltage regulator tolerances and power-on sequencing conditions it may not be possible to guarantee that VDDL never exceeds VDDH. Therefore there is a need for separate N-wells for VDDL and VDDH domains within the level shifter, connected to different voltages.

N-wells at different potentials have specific (and larger) spacing rules than N-wells at the same potential. Because the level shifter cells have to abut with arbitrary standard cells, they must present a standard (VDDH) well at the periphery. Internally a local "hot" VDDL N-well with large spacing rules around the P-MOS transistors of the input buffers is required. This makes the final cell larger than the internal transistor count would suggest.

Therefore the low-to-high level shifters may be multiple-cell row height to facilitate the multiple power well connections and to comply with the well separation design-rules. As a result these cells require careful placement to minimize area. Specialized level shifter cells can also be built which allow the wells to be connected by abutment. This approach avoids the overhead of complete well separation within every level shifter cell but requires specialized placement and EDA tooling.

An example level up-shifter layout is shown in the figure below. To handle the VDD N-well isolation cleanly the cell in this case is built as a triple height cell. This may appear excessive, but it allows completely flexible placement in the high voltage domain. It can be vertically flipped depending on whether the "base" row is a VDD or VSS rail. The fact that there is a large amount of unused space around the well separation is acceptable in this case as it does not require special placement rules and scripting.

Figure 12-7   Example Layout of a Low to High Level Shifter

The advantage of managing the layout of the N-wells internally is that a clean standard cell is presented to the implementation tools.

The "up-shift" design can be extended to provide isolation as well. Figure 12-8 shows an example of a level shifter + isolation cell. When driven low, the "ISOLN" clamp control (controlled from the VDDH domain) effectively turns off the up-shifting amplifier and clamps the level shifter output to zero. The VDDL supply can then be turned off and the buffer outputs can float without causing any crowbar currents.

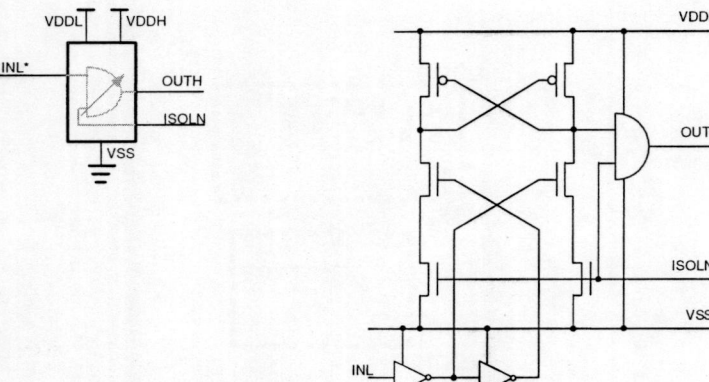

Figure 12-8   Level Shifter Plus Isolation Cell

Incorporating the isolation control within the level shifter can simplify implementation when power gating blocks that operate at different voltages.

12.4 Memories

In most cases, memories are generated from memory compilers. In some special cases, however, optimized memory instances may be built for specific power/performance sensitive applications.

Memory compilers can generate a variety of memory architectures:

- Single or multi-ported RAMs
- RAM arrays or register file architectures
- Performance-optimized memories
- Area-optimized memories
- Power-optimized memories

The performance and area trade-offs are largely due the characteristics of the bit cell and to the underlying banking architecture of the memory. Small banks can be decoded and accessed faster, but the more banks the larger the area.

RAMs can also be designed with high or low threshold voltage transistors. Small cache memories for example are typically tuned for performance and the extra

dynamic and leakage power has to be tolerated. Other on-chip memories however may better be implemented in high-V_T transistors, at least in the periphery circuits, in order to keep the power to a minimum.

12.4.1 RAMs for Multi-Voltage Power Gated Designs

In 90nm geometries and below, RAMs typically have little or no voltage headroom. They must be run at full voltage to meet their timing specification.

In voltage scaling designs, we often run some standard cell logic blocks at voltages lower than full voltage to reduce power. Thus, in multi-voltage designs it is often necessary to level shift up the inputs to the RAM and shift down the RAM outputs.

If power gating or external power rail switching of the logic is supported then to retain the contents of the RAM correctly we need to clamp the inputs to the RAM as well.

Memories are often on the critical timing path of any design, and closing timing is challenging when clocks and inputs are level shifted and clamped. For this reason we would like to place the level shifters and clamps as close to the memory as possible. This avoids any differential path delay between the RAM clock and the RAM control, address and data inputs.

Figure 12-9 shows an integrated multi-voltage RAM interface. In this case the level shifters and clamps are part of the RAM. The level shifters, clamps, and memory are all characterized as a single unit.

The active low isolation clamp signal shown in this example, ISOLN, needs to be driven by a buffer that is always on to ensure it does not float and corrupt the RAM contents. Other inputs may be shut down once the interface is isolated.

If the RAM compiler does not support the generation of interface layers that include level shifting and isolation, then the best alternative is to build and characterize each instance of the RAM with its own discrete level shifters and clamps as a new component.

When several RAMS share interface signals, it can be tempting to group them together and share the level shifters and isolation cells. Unfortunately this increases the distance between the shifters/isolation cells and the RAMs. This in turn results in buffering and interconnects on the far side of the level shifters, making timing closure and clock tree balancing a challenge.

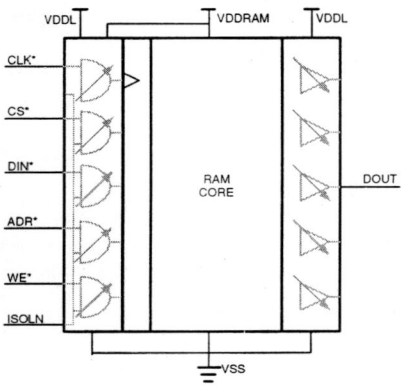

Figure 12-9   Multi-Voltage RAM Interface

12.4.2 Memories and Retention

There are several techniques for lowering the static power of memories when the logic around them is powered down. This is discussed in Chapter 13.

12.5 Power Gating Strategies and Structures

There are several different approaches for gating the power rail. The two most common are:

- "MT-CMOS" – Multi-Threshold CMOS (high V_T switches)
- "MV-CMOS" – Multi-Voltage CMOS (low V_T switches)

MT-CMOS consists of using high V_T switches to turn power off. It is addressed in detail in this chapter, and is simply referred to as Power-Gating.

MV-CMOS consists of using low V_T switches to turn power off. In order to reduce the leakage through these switches during power down, the gate of the switch is driven above VDD (for header switches) or below VSS (for footer switches). The challenge for MV-CMOS is that during power gating, when the transistor is shut off, the control voltage must be outside the VDD and VSS rails. For an NMOS footer cell that switches VSS, the SLEEP control signal must be more negative than VSS. For a PMOS header cell that switches VDD, the SLEEP control signal must be more positive than VDD.

For this reason, MV-CMOS is complex to support. There is a requirement for additional voltage rails, and this may require on-chip charge pumps or external low current supplies. The biggest drawback of this approach is the requirement for non-logic supply rails and special sleep control networks. MV-CMOS is rarely used in commercial designs.

For the rest of the book, power gating will refer exclusively to MT-CMOS power gating.

12.5.1 Power Gating Structures

We begin the discussion of power gating structures by revisiting a question raised in a previous chapter: why use coarse grain power gating rather than fine grain power gating.

Fine-Grain Power Gating

In the fine grain style power gating designs, a sleep transistor is inserted into every standard cell. Cells with embedded sleep transistors are often called MTCMOS (Multi-Threshold CMOS) cells. Figure 12-10 shows two examples of an MTCMOS AND gate, one with a footer, the other with a header switch.

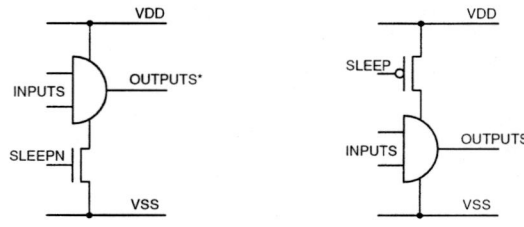

Figure 12-10   Fine Grain Cells

A power gating control signal "SLEEP" (or "SLEEPN") controls the sleep transistor to switch on and off the power supply to the cell.

Since the power switch must supply worst case current required by the cell, it has to be quite large not to impact performance. In fact, the switch often can be several times as large as the rest of the cell.

In order to keep this area overhead to a minimum, fine grain power gates are usually implemented as "footer" switches, switching VSS rather than VDD. This is because NMOS transistors have a lower on-resistance than PMOS and so will be smaller.

Even using footer switches, the area overhead of each cell is quite large (often 2x-4x the size of the original cell).

To further reduce the area overhead, most designs only power gate the high leakage, low threshold cells.

Mixing power-gated and always-on cells creates another problem. When the power to the gated cell is turned off, the output will float, and may float to the threshold voltage. If this output is connected to a cell that is still powered up, then crowbar current could result. For this reason, a weak pull-up/down transistor is often added to clamp the cell output to a known state during power down. The pull-up/down transistor remains in OFF state in normal operation.

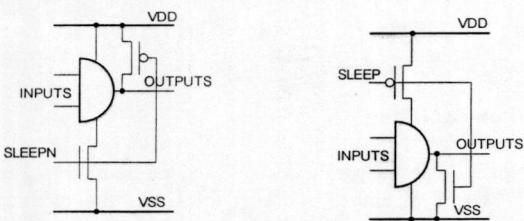

Figure 12-11   Fine Grain AND with Pull-Up

Fine-grain power gating has several advantages:

- It is not sensitive to ground noise injection because the virtual power nets are short and hidden in the cells.
- It has small wake-up latency and small in-rush current at wake-up, due to the small capacitance in the virtual power.
- The built-in clamp transistors keep all outputs at a known state which effectively eliminates crowbar current in the CMOS cells during the wake-up period.
- the timing impact of the IR drop across the switch and the behavior of the clamp are easy to characterize as they are contained within the cell
- It can be synthesized and analyzed by conventional ASIC tools and flows, because the MTCMOS library cells can be modeled and characterized in a same way as standard cells. Only the added "sleep" signal pin needs special attention in the design flow. The delay and IR drop effects of the built-in sleep transistors can be accurately accounted for in the cell characterization.

However, fine-grain power gating also has several disadvantages:

- It introduces significant area penalty due to the addition of a sleep transistor in every cell. The cell area increase can be up to 3x for keeping low IR drop in the cells and acceptable performance degradation.

- It requires a specially designed MTCMOS cell library.
- It needs significant buffering and routing resources to distribute sleep control signal to all the cells in a design.

Coarse-Grain Power Gating

In the coarse grain power gating designs, the sleep transistors are connected in parallel between the permanent power and the virtual power networks. As in fine grain power gating, the sleep transistors can be either header switches (switching VDD) or footer switches (switching VSS). Figure 12-12 shows an example of coarse grain power gating with footer switches controlled by a common sleep control signal: "SLEEPN."

Coarse grain power gating has following advantages:

- Because the sleep transistors can share charge or discharge current in the design, it is less sensitive to PVT (process, voltage, temperature) variation in the sleep transistors and introduces less IR drop variations than the fine-grain power gating design.
- It has significantly smaller area overhead than the fine-grain power gating. The number of the sleep transistors can be optimally tuned for design specific IR drop and speed targets.
- It can utilize existing standard cell libraries. Only a few special cells, such as sleep transistors, isolation cells, and retention registers need to be added to the libraries.

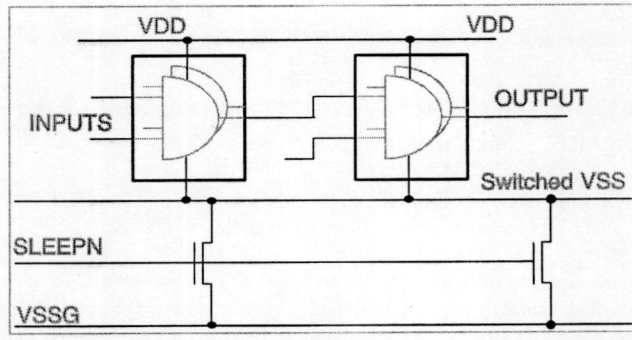

Figure 12-12   Coarse Grain Power Gating

However, the coarse grain power gating design has its disadvantages:

- It requires a complex power network including the permanent power network, the sleep transistors, and the virtual power network. Consequently, the power network synthesis becomes challenging and requires rigorous static and dynamic IR drop analysis.
- It requires wake-up in-rush current control to prevent power supply noise and possible data corruption.
- It has longer wake-up latency, due to the time needed to charge up the large virtual power network.
- It adds complexity to STA and power analysis because cell delay depends on IR drop on the sleep transistors. This inter-dependency of timing, IR drop, and power requires simultaneous analysis of all three to be accurate.
- It imposes special logic and physical constraints in logic and physical synthesis, resulting in more complex design methodologies and flows.

12.5.2   Recommendations – Coarse Grain vs. Fine Grain

- Most design teams have found the area penalty of fine grain power gating to be prohibitive. As a result, most power-gating designs use the coarse grain power gating style. In the remainder of the chapter (and this book), we focus on coarse grain power gating.

12.6   Power Gating Cells

A standard cell library that supports power gating should include both header and footer power switches. A range of switch sizes and strengths enables a variety of different switch network designs.

Figure 12-13 shows the abstract schematic of a footer and header switch. Here VSS is the switched VSS and VSSG is the always on VSS. VDD is the switched VDD and VDDG is the always on VDD.

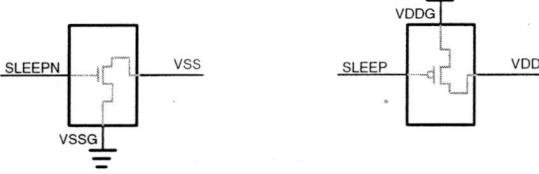

Figure 12-13   Footer and Header Switches

A switch cell is physically made up of a number of parallel switch transistors which are carefully sized to maximize the on-current to off-current ratio (Ion/Ioff).

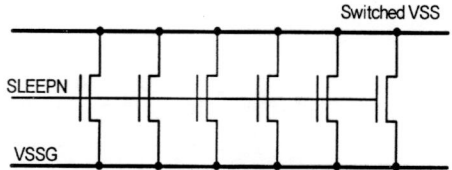

Figure 12-14   Parallel Transistors Make Up the Switch

The layout for an example footer switch cell is shown in Figure 12-15. The SLEEPN control input port is shown on the left. The global VSSG rail must be connected to the interleaved contacts spread across the center of the cell. The switched virtual ground rail VSS appears on the bottom track to connect to standard cell ground track. The VDD rail at the top is simply routed through as a standard cell power track, and will normally be connected to the always on VDDG supply grid in the footer-switched system.

Rather than trying to produce one monster switch structure, a range of at least two or three switch sizes or strengths is typically provided. These can then be distributed across the design using a global rail grid and a switched rail grid to maximize current sharing between adjacent switches.

Turning on all such switches instantaneously would potentially cause the supply rails to collapse – corrupting retention state or adjacent powered logic sharing the global rails.

Therefore smaller switches are provided which can support reduced current turn on strategies. For instance, an implementation may use small switches to turn on the switched rail and provide an initial charging current. When the rail hits a certain voltage, the larger switches are then turned on.

The implementation phase handles how many switches are needed and in what topology. Detailed analysis is required to manage the virtual rail IR drop and the turn-on in-rush current. These issues are addressed in more detail in the next chapter.

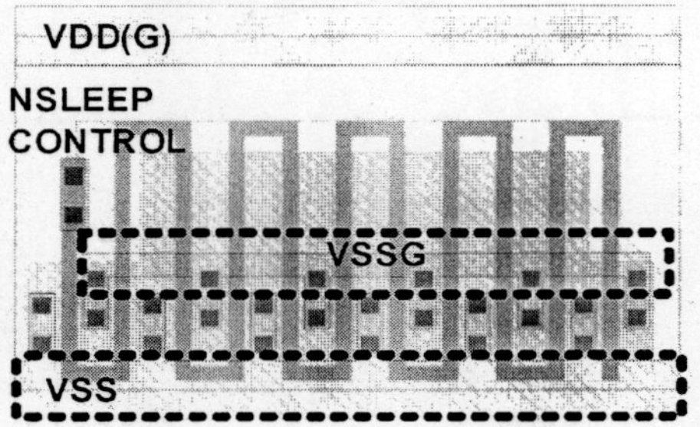

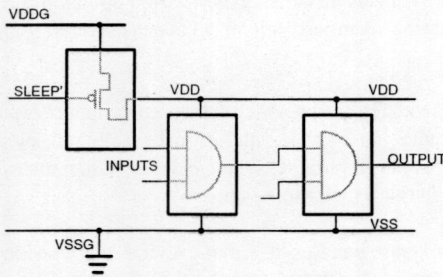

Figure 12-15   Layout of a Footer Cell

12.7   Power Gated Standard Cell Libraries

With coarse-grain power gating, we use ordinary standard library cells. One key challenge is to deal with the effects of the IR drop across the power switch on the timing of these cells.

If header switches are used then the standard cells will be connected as:

Figure 12-16   Header-based Power Gating

Because the IR drop is design and placement specific there are two approaches to dealing with the voltage drop de-rating on timing of power gated standard cells:

- Ignore the timing effects of the IR drop at design time and over-constrain the design target frequency. This can be a reasonable approach during early synthesis runs when we are not modeling the low power aspects of the design.

- Set a realistic IR-voltage drop for the supply rails and work with timing libraries that support voltage de-rating. Then in the back-end analysis check the instance-based de-rating to ensure the IR drop is never exceeded. One can start with a mesh of many big switches and optimize these down to weaker, less leaky switches once the "hot-spots" are satisfactorily dealt with.

Note: the design frequency for power gated designs will realistically be 5%-10% lower than without power gating. Over-ambitious frequency goals will simply result in throwing more and more power switches into the design to try to close timing.

CHAPTER 13

Retention Register Design

For those designs that require fast resumption of operation after wakeup, it is necessary to save the current state of a design before going into sleep mode and to restore the state at wakeup. In this chapter, we describe on-chip retention methods, including retention registers and retention techniques for memory.

13.1 Retention Registers

There are a number of different retention register designs. The ones we discuss here are all variations of a standard scan-testable D-type flip-flop. We describe three kings of retention registers:

- Single Pin Live Slave – uses a single save/restore control pin with minimal changes to the flop itself
- Single Pin Balloon – uses a single save/restore control pin but adds a second slave latch for retention
- Dual Pin Balloon – uses separate save and restore control pins and a second slave latch for retention

13.1.1 Single Pin "Live Slave" Retention Registers

The simplest form of retention register is one in which the underlying master-slave latch structure is adapted to provide a low-leakage mode to maintain the state of the slave latch.

Figure 13-1 shows the conceptual adaptation of the rising-edge clocked scan-register design.

The front-end of the register is a multiplexer. When the scan-enable control "SE" is de-asserted it selects the functional data "D" input; when SE is asserted it selects the scan chain serial input "SI." To provide best setup timing behavior this mux and much of the data path use Low-V_T transistors – indicated in the drawing by a thick bar on the gate.

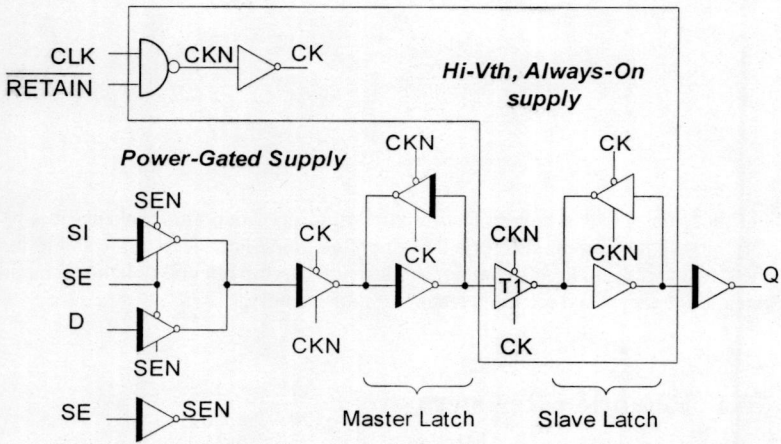

Figure 13-1   Retention Register: "Live Slave"

The "Master" latch samples the input value when the internal clock "CK" is low and the inverted "CKN" is high. The master latch uses Low-V_T transistors to improve set-up and hold characteristics.

The "Slave" latch samples the output of the master latch on the alternate phase of the internal clocks and is followed by an inverting buffer stage. This last inverter is implemented with various transistor sizes to provide different drive-strength versions of the register in the library.

Retention is added by introducing the following features:

- An AND-gate is inserted in the clock circuit to allow the retention signal (NRE-TAIN) to force the clock off during power-gating.
- The slave latch is powered from an always-on supply while the rest of the register is powered by the virtual rail and can be power gated off.

- High-V_T transistors are used in the slave latch, the clock buffers, and the inverter that connects the master latch to the slave latch.

The active-low "NRETAIN" control signal must be carefully controlled to change only when the clock is inactive (low in this case). Also it must be driven by an always-on network of buffers.

In system usage the clock is stopped; then the NRETAIN signal is asserted low; finally power is gated off. NRETAIN is held low while power is gated off.

After the gated power is re-applied, the NRETAIN signal is then be de-asserted (high) before the clock is restarted.

While NRETAIN is asserted, the three-state inverter (labeled T1) turns off, isolating the slave latch from the master latch.

For a register with asynchronous set or reset, the retention control would also need to isolate the register from power-gated reset or set networks that could also corrupt the slave latch signal.

Advantages of the "Live Slave" retention register design include:

- Minimal area impact on the underlying master-slave latch design
- A single signal controls retention

Disadvantages of the "Live Slave" retention register design include:

- Performance impact on the register. The slave latch is implemented in High-V_T to minimize leakage in retention mode. However this impacts the CLK to Q timing path, and the transistor sizing for best leakage has to be sacrificed to balance the compromise in performance.
- The High-V_T gating of the clock increases the hold-time requirements of the input data. Even then, we cannot use minimum leakage transistor sizes because of the number of internal low-V_T clock nodes that have to be driven with good balanced rise and fall times.
- It is only appropriate if the clock can be forced low before restoring state – any "unknowns" in the clock gating circuit would need explicit overrides.

As a library component such a register appears as a component with both always-on and power-gated supply rails, and the extra retention control signal. A generic component that could be used with either header- or footer-switched power gating is shown in Figure 13-2 on page 212:

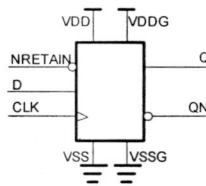

Figure 13-2   Generic Switch Cell

The "global power" VDDG/VSSG supplies the high-V_T internal clock gate and buffering as well as the slave latch. The rest of the register circuitry is power gated from the VDD/VSS rails.

Functional control waveforms are shown Figure 13-3. The clock must be stopped and "parked" inactive with logic 0 value. The retain signal must be asserted (NRETAIN low in this case) before the gated power is turned off. The retain signal must only be de-asserted after the power is restored, and only then may the clock be restarted. The power gating signal (PG_ENABLE) powers down the gated supplies, including the clock tree buffering as shown in the clock waveform.

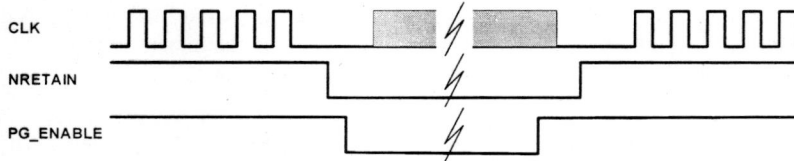

Figure 13-3   Control Waveform for Single Control Retention Register

13.1.2   Dual Control Signal "Balloon" Retention Register

An alternative design of retention that maintains the underlying register speed and clock-to-output performance is the balloon retention register. It adds a weak high-V_T latch to a standard D-type flip-flop. This third latch (sometimes called a shadow latch or "Balloon" latch) is connected to an always-on power supply and holds the register state while the leaky master-slave register latches are powered down.

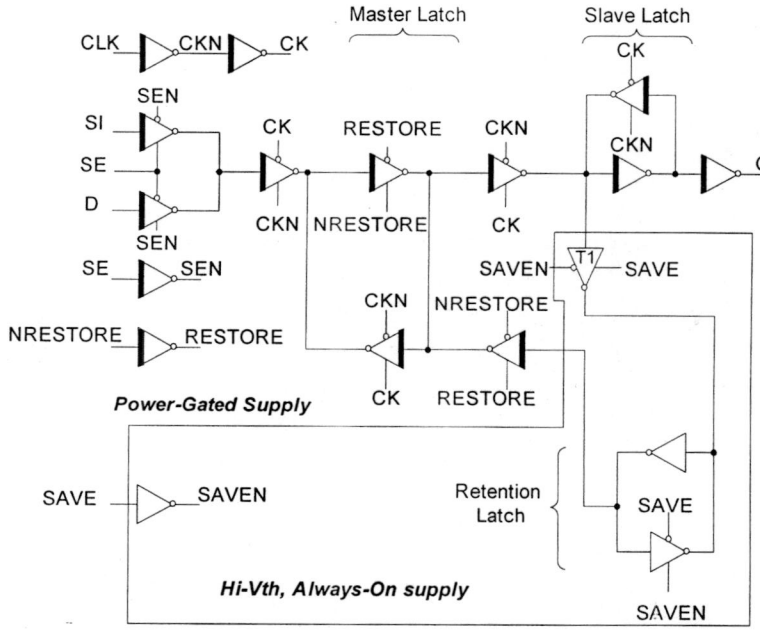

Figure 13-4   Dual Control Balloon Register

As shown in Figure 13-4, the basic register design is implemented in low V_T transistors. A low-leakage, high V_T retention latch is added in an always-on supply voltage region.

Two control signals have been added to the register. An always-on signal "SAVE" is used to control the sampling of data into the retention latch. When SAVE is asserted, the state of the slave latch is copied into the retention latch. This data is then maintained whenever the SAVE signal is de-asserted. In system usage the clock should be stopped and the SAVE signal pulsed before power gating to preserve the state of the register.

An active-low restore signal, NRESTORE provides the control to force the state of the register to that of the retention latch value. In system usage the gated power must be restored and safely stabilized before the NRESTORE signal is pulsed (low) to set the state of the register back to the retained value.

A rather complex example design for the register restore paths is shown in the figure. This design allows the retention register to restore the retained value regardless of the state of the clock. If the clock is low and the master latch is open and sampling input data, the retained value is forced into the slave latch. To prevent contention when restoring, if the clock is high the retention latch value is forced into the master latch, which then propagates to the slave when the clock goes low.

Advantages of this Save/Restore control style of retention register design

- Minimal leakage power. The retention latch and control signal can be minimal transistor sizing.
- Minimal performance impact compared to the "Live-Slave" design, although there is some minor internal loading on the output of the master latch and the input of the slave latch where the save/restore control transistors are added.
- Can be built to be independent of clock phase on restore, which can be valuable in designs with complex clock gating.

Disadvantages of this Save/Restore control style retention register design

- Area impact on the underlying master-slave latch design with addition of a third latch – even though only a small transistor structure is added.
- Using two control signals (SAVE and NRESTORE) requires two buffer networks, only one of which (NRESTORE) can be power gated – which adds some area impact.
- The dual control signaling adds some complexity to the system design and some buffer tree area.

Figure 13-5 shows the library component view of such a retention register with two independent, asynchronous, pulse-style signals to control the save and restore function. The "global power" VDDG/VSSG supplies the high V_T save control buffer and retention latch. The rest of the register circuitry is powered from the (switched) VDD/VSS rails.

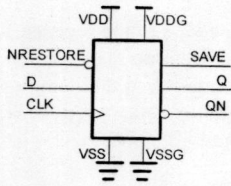

Figure 13-5   Dual Control Retention Register

Functional control waveforms are shown Figure 13-6. The clock may be stopped in either phase. The state save signal must be pulsed active (SAVE high in this case) before the gated power is turned off. The state restore signal must only be pulsed after

the power is restored, and then the clock may be restarted. The power gating may power off the clock tree buffering and the restore control buffer network as shown by the unknown logic level in the waveforms.

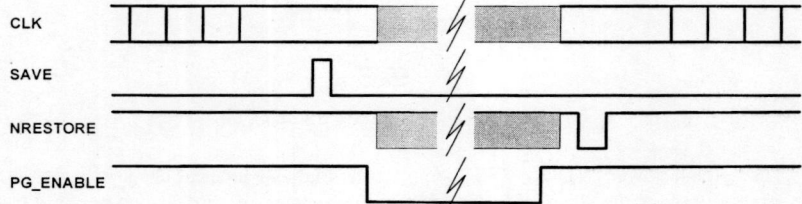

Figure 13-6   Control Waveform for Dual Control Retention Register

Precedence of Retention and Asynchronous Resets and Presets

To minimize leakage we would like to power down the high fan-out reset and set networks as part of the system power gating. Once powered down, these nets can float, so these ports on the retention flops may float to non-logic levels during power gating entry and exit.

By ensuring that the balloon latch is controlled only by the state save control network, any asynchronous sets and resets only operate on the master/slave latches. It is then a system level sequencing requirement to assert the restore control only after resets and presets have been powered up and put in the appropriate state.

13.1.3   Single Control Signal "Balloon" Retention Register

An enhancement to the SAVE/RESTORE style of retention register described above is to use a single control signal for both save and restore. That is, state is saved on one edge of the control signal and restored on the other edge.

Figure 13-7 shows the conceptual schematic for a balloon-style retention register with a single state retain control – an active-low NRETAIN signal in this example.

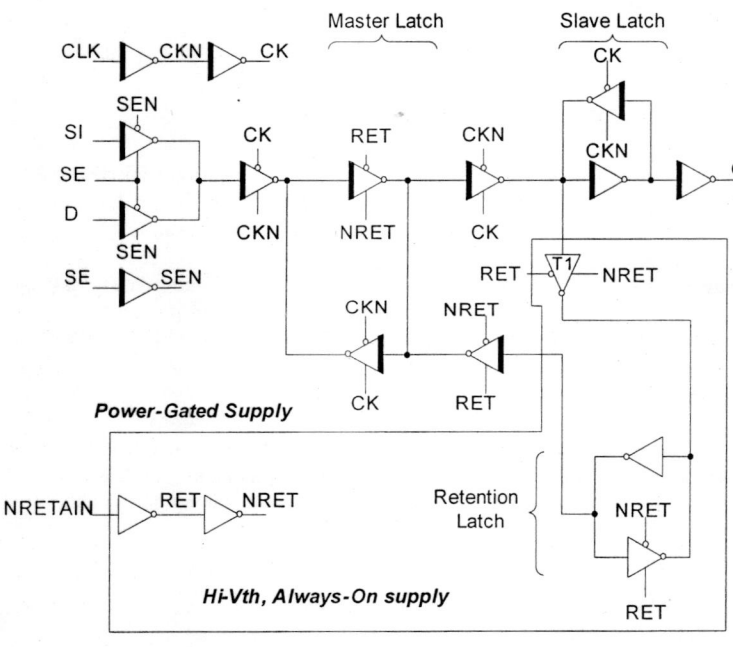

Figure 13-7   Single Control Balloon Register

The NRETAIN signal must be driven by an always-on control network and is implemented in high V_T to minimize leakage during power gating. The structure is very similar to the two-control signal register style described above, but in this case the retention latch samples the slave data whenever the NRETAIN signal is de-asserted high, and holds the state whenever the NRETAIN signal is asserted low.

This design has slightly higher dynamic power because the extra capacitive nodes of the retention latch must be driven whenever the slave latch changes state. Because the retention latch is designed with small weak transistors to minimize leakage power, this is a small proportion of the main master-slave register dynamic power. This design however dispenses with the requirement for two control networks.

Similar to the preceding dual-control register, a rather complex example design for the register restore paths is shown in the figure to allow this retention register to restore the retained value regardless of the state of the clock. If the clock is low and the master latch is open and sampling input data the retained value is forced into the slave latch. If the clock is high however the retention latch value is forced into the master latch, which then propagates to the slave when the clock goes low.

Advantages of this Save/Restore control style of retention register design include:

- Minimal leakage power. The retention latch and control signal can be minimal transistor sizing.
- Minimal performance impact compared to the "Live-Slave" design although there is some minor internal loading on the output of the master latch and the input of the slave latch where the save/restore control transistors are added.
- Can be built to be independent of clock phase on restore, which can be valuable in designs with complex clock gating.
- Single control network compared to the dual-control style of retention register. This saves system-level dynamic power because only one, rather than two, sets of buffers is required to distribute retention control.

Disadvantages of this Save/Restore control style retention register design

- Area impact compared to the "Live-Slave" design due to the addition of a third latch – even though only weak small transistor structure added.
- Slightly higher dynamic power than the dual control balloon design, as the retention latch transitions every time the slave latch value changes.

Figure 13-8 shows the library component view with the single asynchronous retention control signal used to provide both the state save and restore function. The "global power" VDDG/VSSG supplies the high V_T buffer and retention latch, the rest of the higher performance register circuitry can be power gated from the VDD/VSS rails.

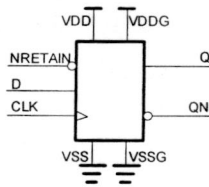

Figure 13-8   Single Control Retention Register

Functional control waveforms are shown Figure 13-9. The clock may be stopped in either phase. The retain signal must be asserted (NRETAIN low in this case) before the gated power is turned off, and only de-asserted once the power is safely restored, before the clock may be restarted. The power gating may power off the clock tree buffering as indicated by the unknown logic level in the waveforms.

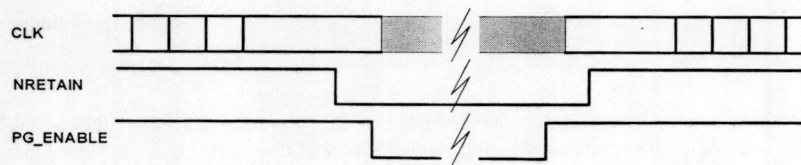

Figure 13-9   Control Waveform for Single Control Balloon Register

13.1.4   Retention Register: Relative layout

Figure 13-10 shows an example of a standard scan-testable D-type register. The "Live-Slave" version of this would be a little larger to include the clock gating control and high V_T internal clock phase buffering with an area of high V_T implant.

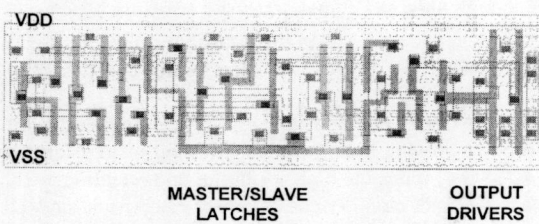

Figure 13-10   Standard Scan D-type Register

Figure 13-11 shows an example of a "balloon style" retention style scan-testable D-type register. The high V_T implant area for the retention latch and control buffering is clearly visible as extra area for the cell implementation. For higher drive-strength register versions the retention area becomes a smaller proportion of the total register cell area.

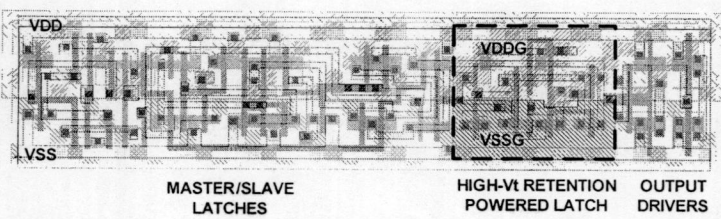

Figure 13-11   Balloon Register

As show, the area overhead for adding retention to a basic, low drive-strength register is about 30-40%.

13.2   Memory Retention Methods

FIFOs' are often flushed before going into sleep and caches are initialized after the wakeup in the power-gating designs. In these cases, we can power down the memories to save static power and tolerate the loss of the memory data.

However, for high performance designs which require minimum latency at wakeup, the contents of on-chip memory need to be retained during power gating. Various memory retention methods have been developed. The principle of these methods is to reduce leakage as much as possible during power gating without corrupting the data in the SRAM.

It is not practical to introduce a retention circuit like those discussed above into the SRAM cells. Any such circuit will cause an unacceptable area increase. Instead the VDD retention and the source bias retention methods are among those methods which are frequently used in power-gating designs.

13.2.1   VDD Retention Method

In this method, a separate VDD power supply is provided to the memory. In normal operation, VDD is provided at the normal supply voltage. In sleep mode, VDD is reduced to 0.5-0.6V to reduce memory power consumption while maintaining memory contents.

This method is simple to implement. No memory circuit change is required. However, it requires a dedicated, switchable power supply to the memories.

13.2.2   Source-diode Biasing Method

The principle of the source biasing SRAM retention method is to apply reverse body bias for further leakage reduction after lowering the SRAM operation voltage. The source-diode biasing method is the simplest implementation of the source biasing method. In this method, a diode is inserted in the source supply to the SRAM cell array and is controlled by a switch as shown in Figure 13-12.

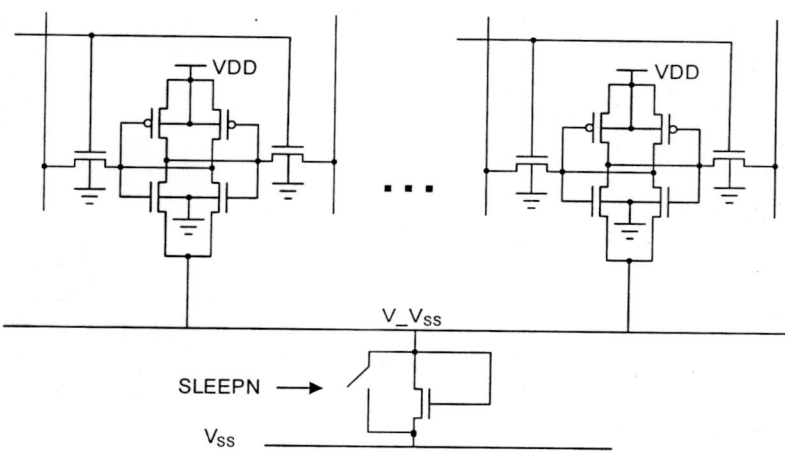

Figure 13-12   Source-diode Biasing SRAM Retention

In normal operations, the control signal SLEEPN is de-asserted, so that the switch closed, bypassing the diode. In this case V_Vss = Vss.

In sleep mode, SLEEPN is asserted, which opens the switch. The source supply of the SRAM cell array is now through the diode. The built-in threshold (V_T) of the diode raises the voltage of V_Vss above Vss (ground). The SRAM operation voltage is reduced by V_T and hence the leakage reduced, just as in the VDD retention method.

Note that the substrate of the NMOS transistors in the SRAM array is connected to ground (Vss), which is now lower than the source bias at V_Vss. Thus, the voltage rise in V_Vss also applies a reverse body bias to the NMOS transistors cells. The reverse body bias further reduces sub-threshold leakage current in the SRAM array cells in sleep mode.

The same principle also applies to the PMOS transistors in the SRAM array cells where a diode can be inserted in the VDD supply. However, this is not applicable to sub-90nm SRAM designs where VDD is scaled to sub-1.2V, because a 2x V_T (on both diodes) voltage reduction results in an SRAM operating voltage too low to retain data.

So either PMOS or NMOS biasing can be used in sub-90nm designs. Since PMOS transistors in the SRAM cells are less leaky than the NMOS ones, NMOS source biasing is typically used in these designs.

The bias control switch is commonly implemented by a high V_T NMOS sleep transistor and the diode is made by an NMOS transistor with gate and drain connected.

The advantage of the source-diode biasing method is its easy implementation. No bias supply is required.

However, it is difficult to obtain the optimal reverse body bias for maximum leakage reduction while retaining data. The reverse body bias is determined by the V_T of the diode and hence is fixed for a given process.

13.2.3   Source Biasing Method

The noise margin in SRAM cells shrinks with VDD scaling. This requires optimal SRAM operating voltage and source bias in sub-1V designs. One solution is to provide a dedicated source bias supply to the SRAM cells to replace the diode. This is illustrated in Figure 13-13.

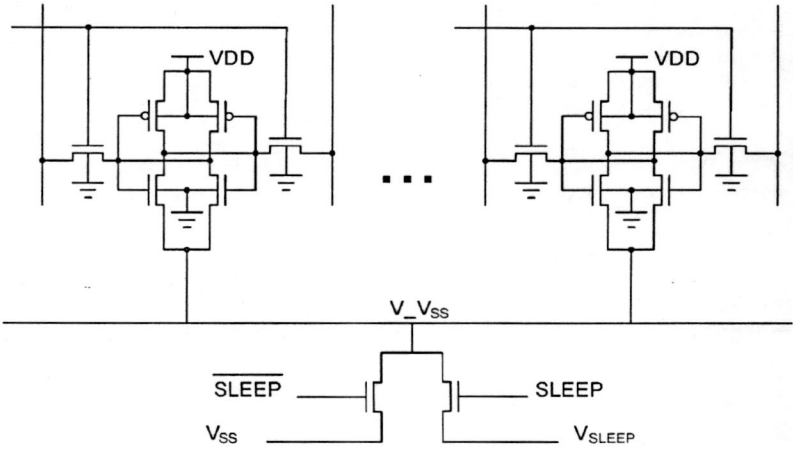

Figure 13-13   Source biasing SRAM retention

The disadvantage of this method is the requirement of the separate biasing supply. Since SRAM consumes only leakage current in the retention mode and the biasing is in the power supply range, it is possible to generate the biasing supply on chip with a simple design such as a voltage divider.

13.2.4   Retention Latency Reduction Methods

For those applications where memories are not frequently accessed but access latency needs to be short, the memory retention methods discussed above can be combined with retention latency reduction techniques. Two such techniques are described below.

Block-based retention and wakeup

In this method, a memory is divided into a number of small size blocks or banks. Each block can be individually controlled to be in retention or function mode. All memory blocks are normally in low-leakage retention mode. When a memory read or write operation is requested, the address decoder selects the access block which is then switched from retention to function mode, ready for data access.

After the data access, the block immediately returns to retention mode for power saving. The retention latency, i.e. the time needed to waken a block into fully function state, is shortened by reducing the size of the virtual VSS network of the block. The smaller the block, then the quicker the discharge of the block's virtual VSS and hence the shorter the wakeup latency. However, if the block becomes too small, then the area and power overhead in the column sense-amps could overwhelm the benefit of the leakage power saving in retention.

We recommend exploring various memory banking strategies based on the design specific memories, access frequency, and retention latency requirements.

Row-based retention and wakeup

The row-based retention method addresses the issue of significant column sense-amp overhead in the block-based retention method. In the row-based method, the size of the memory block is determined by the overall considerations of memory power consumption and access time. This usually results in medium size memory blocks.

Unlike the block-based retention method, the V_VSS nets in the row-based method are distributed per row. Each row is individually controlled by its V_VSS net in either retention or function mode. Taking the advantage of the memory access control sequence where memory address is put on bus before data, the method uses the SRAM word address to activate only the row which is required for the current memory access. The rest of the SRAM array cells in the other rows remain in retention mode. The row-based retention method reduces latency because the small size of a row allows fast power and power down. Hence we do not need to split the memory into many arbitrary, small blocks which would incur significantly area and power penalties in column circuitry.

To reduce the complexity and overhead of the control circuitry, the row-based retention and wakeup control can be extended into a row-group-based method where a group of rows, instead of a single row, is controlled during retention and wakeup.

CHAPTER 14

Design of the Power Switching Network

Power gating is the most effective method for reducing leakage power in standby or sleep mode. However, this method comes with overhead such as the silicon area taken by the sleep transistors, the routing resources for permanent and virtual power networks, and the complex power-gating design and implementation processes which impact design risk and schedule.

Besides the overhead, power gating introduces power integrity issues such as IR drop on the sleep transistors and ground bounce caused by in-rush wakeup current. It also introduces wakeup latency, the time needed to restore full power for normal operation. All these issues must be addressed during the implementation of power gating designs.

In this chapter, we discuss:

- power gating implementation styles
- wakeup in-rush current control
- wakeup and sleep latency reduction
- sleep transistor power network synthesis

14.1 Ring vs. Grid Style

Coarse grain power gating can be implemented in either a ring or a grid style power network.

With ring based switching, we place the switches externally to the power gated block effectively encapsulating the block with a ring of switches.

In the grid style implementation, the sleep transistors are distributed throughout the power gated region.

14.1.1   Ring Style Implementation

Figure 14-1 shows an example of a ring implementation. A ring of VDD surrounds the power gated block. A ring of switches connects VDD to a switched or virtual VDD (VVDD) power mesh that covers the power gated block.

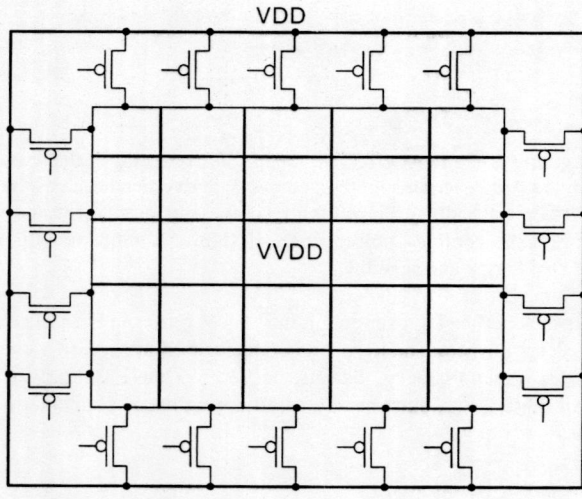

Figure 14-1   Ring Style Sleep Transistor Implementation

Note that a ring style switching network is the only style that can be used to power gate an existing hard block. The switches and VDD can be added outside the hard block, and the hard block's VDD power mesh can now be used for VVDD.

The ring style sleep transistor implementation has following advantages:

- It has a less complex power plan than the grid style because of the separation of the permanent power network and the virtual power network. Moreover, the sleep transistors are confined in the regions around the virtual power network and are not mixed with other logic cells.

- It has little negative impact on placement and routing in the standard cell area because neither the permanent power nets nor the sleep transistors are in the areas where the logic cells are placed and signals are routed. Those special cells, such as isolation cells and always-on buffers that require the permanent power supply can be placed around the power domain areas.

The ring approach can be a good option for small blocks of logic where the voltage drop across the switch transistors and VVDD mesh can be managed. However, for larger blocks of logic, managing the voltage drop with a ring based approach can be difficult.

This approach can also be useful for legacy (hard) IP or optimized logic blocks where re-implementing those blocks would be costly.

However, the ring style has some significant disadvantages:

- It is does not support retention registers, since these require access to the always on supply.

- A ring based approach can add significant extra area cost compared to a grid approach.

14.1.2   Grid Style Implementation

In the grid style implementation, the sleep transistors are distributed throughout the power gated region. They form a grid to connect permanent power network and virtual power networks, as shown in Figure 14-2 on page 228.

VDD

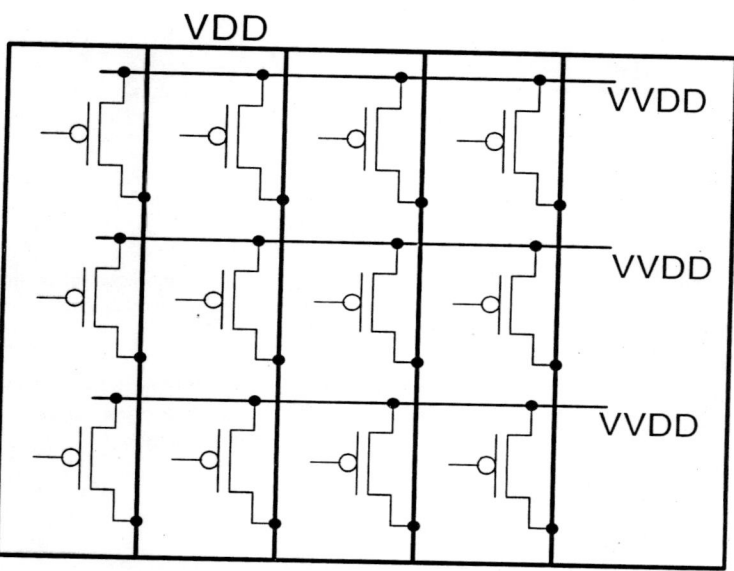

Figure 14-2   Grid Style Sleep Transistor Implementations

The grid style sleep transistor implementation has following advantages:

- With the sleep transistors distributed across the power-down domain, the switches in a grid network do not have to drive the virtual supply for the long distances incurred in the ring-style distribution. As a result, the virtual power network can be implemented in low metal layers. The wide straps used in the ring-style implementation are not required.

- It requires fewer sleep transistors than the ring-style implementation to achieve the same IR drop target. This is once again due to the fact that the transistors are distributed and do not have to drive long metal interconnect.

- The permanent power supply is available across the power-down domain areas. Consequently, special cells, such as retention registers and always-on buffers, which require the permanent power supply, can be connected to the permanent power network in the power-down areas.

- It provides somewhat better trickle charge distribution for management of in-rush current.

- It has less impact on the area of a power gated block. Typically the utilization of any block is less than 100%, so there are places where switch cells can be placed without increasing the area of the block.

The drawback of the grid style implementation is its impact on standard cell routing and physical synthesis. Since the sleep transistors are placed in the standard cell area, their placement and routing constraints affect cell placement and net routing.

Also, by distributing the switching function across the design we have added complexity to power routing. We now need to distribute always-on power to the switches as well as the retention registers, isolation cells, and always-on buffers.

14.1.3   Row and Column Grids

With a distributed approach, we are placing the switches internally to the power down block. Many choices are available for the distribution of these switches but they are all basically some type of sparse array, where the switches are placed in an array across the design, each switch separated horizontally by some distance (x) and vertically by some distance (y). When y is minimal we have a column structure and when x is minimal we have a row based structure.

A column based topology employs columns of switch cells spaced evenly across the switched design. These switch cells effectively switch the power rail to each segment of a standard cell row and provide very fine control over the switching function. Each power switch only has to provide power to a small segment of the standard cell row thereby minimizing any potential voltage drop problem.

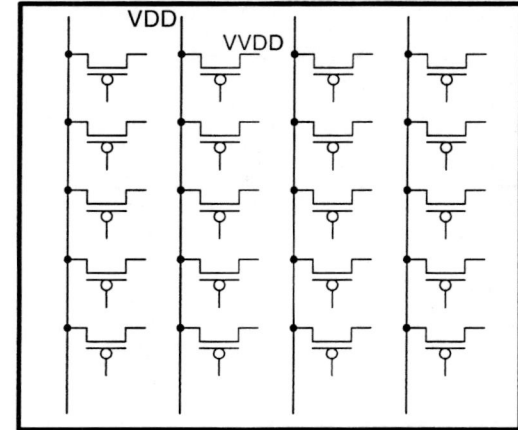

Figure 14-3   Column Based Switching

Of course, distributing the switch cells in columns across the design will impact the placement optimization. These columns of switch cells act as regularly spaced placement blockages, limiting the flexibility of the standard cell placer.

A row based topology may be a more optimal solution for distributed switching since the potential impact on the placement engine is limited as all switch cells are in a single row.

The row approach will take away of a row of standard cells from the placer, but should not impact the placement of logic in other rows of the design.

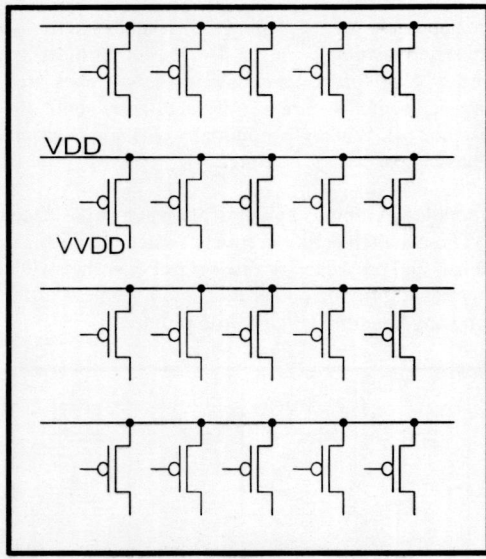

Figure 14-4   Row Based Switching

However, the row based approach can impact routing resources in lower layer metal. This problem can be avoided in the column based approach; the lower layer power straps can be routed in metal 2 directly above the power switches with minimal impact to routing resource.

All of these various topologies have various advantages and disadvantages and it is important to note that the best choice depends on:

- The design being implemented
- The library being used and the type of switches available

- The technology being targeted and its specific leakage characteristics
- The performance and power goals for the design
- The use of legacy or highly optimized IP

Regardless of the initial placement of the switches, modern implementation tools can optimize both placement and sizing of the switches to reduce IR drop and improve timing.

14.1.4   Hybrid Style Implementation

In the hybrid style power gating designs, the grid style is implemented at the top-level and the ring style implementation is applied to certain power-gated hard macros and/or power domain blocks which do not have retention cells.

The hybrid style combines the advantages of the ring and grid style implementations. It is helpful in the cases where a power-gated block has potential congestion and routability issues in the grid style power gating implementation.

However, power planning becomes more complex in the hybrid style due to the mixed ring and grid style power distributions.

14.1.5   Recommendations - Ring vs. Grid Style

- For those designs which implement retention cells, the grid-style implementation is the right choice.
- If there are no retention registers in a design, the choice of grid- vs. ring-style implementations depends on area budget and need of permanent power supply in the power-down areas for always-on buffers. The ring-style should be considered when area is not the main concern and there is no need for a permanent power supply in the power-down areas.
- For those designs which have power-gated hard macros, or blocks that do not contain retention logic, the hybrid style is a good choice, provided that power planning complexity does not become an issue.
- In the grid-style implementations, use wide straps in the permanent power network to reduce the IR drop. The virtual power network should be implemented at metal 1 and metal 2 layers with narrow straps sufficient to drive local logic cells and satisfy the IR drop target. It is worth noting that the total IR drop is composed of the IR drops in the permanent power network, the sleep transistors and the virtual power network. It is preferable to minimize IR drop in the permanent power network so as to make it easier to achieve to total IR drop specification with fewer sleep transistors.

- It may be necessary to experiment with various topologies and compare and contrast the results from the various approaches to determine which is most suitable for the design in question.
- Be sure to consider the power routing impacts for the switch topology chosen.

14.2 Header vs. Footer Switch

A header switch uses a high V_T pMOS transistor to control VDD; a footer switch uses a high V_T nMOS transistor to control VSS. Either a header- or footer-based switching fabric can be used in a power gating design. The key issues affecting this design decision are area cost, IR drop constraints, and system architectural issues. System architectural issues were discussed in Chapter 7. In the following sections we discuss the area and performance trade-offs for headers and footers.

14.2.1 Switch Efficiency Considerations

The sleep transistor switch efficiency is defined as the ratio of drain current in the ON and OFF states (Ion/Ioff). We would like to maximize Ion/Ioff to achieve high drive in normal operation and low leakage in sleep mode.

Although a pMOS transistor is less leaky than an nMOS transistor of the same size, the total leakage in the switch fabric is mainly determined by the switch efficiency. This is because the total leakage also depends on the total number of sleep transistors required to produce the required Ion.

Figure 14-5 and Figure 14-6 show two switch efficiency curves for a 90nm high V_T pMOS transistor and a high V_T nMOS transistor. The simulations were done with normal body bias.

In both cases, the maximum switch efficiency occurs at a gate length of 140nm. The narrow channel effect causes the switch efficiency to change significantly when the gate width is smaller than 0.8um. Moving away from the small width (to avoid problems with process variation), the switch efficiency at 2.2um gate width is 15,000 in the pMOS transistor and 40,000 in the nMOS transistor. This indicates that for a same drive current, the header switches would result in 2.67 more total leakage than the footer switches.

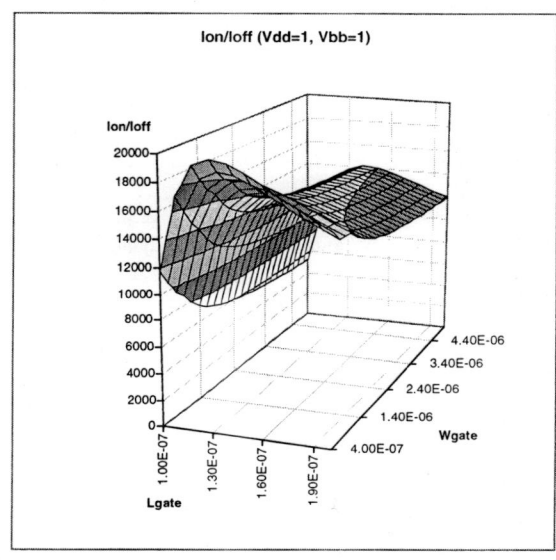

Figure 14-5   90nm High V_T pMOS Switch Efficiency at Normal Body Bias

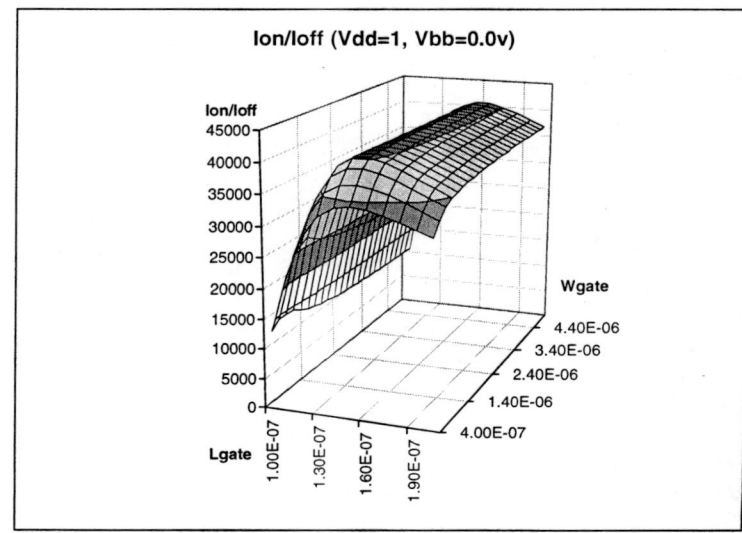

Figure 14-6   90nm High V_T nMOS Switch Efficiency at Normal Body Bias

14.2.2   Area Efficiency Consideration and L/W Choice

The area efficiency of the switching fabric depends on the size (L*W) and layout implementation of the sleep transistors. The optimal L is determined by the switch efficiency and can be obtained from the switch efficiency curves generated from SPICE analysis. Once L is defined, the area efficiency is mainly determined by the transistor width W and by layout implementation.

The switch efficiency decreases with the increase of W in pMOS transistor, as shown by the solid curve in Figure 14-7, due to the narrow channel effect, which affects Ioff more significantly than Ion. Therefore, it is preferable to choose small W for high switch efficiency.

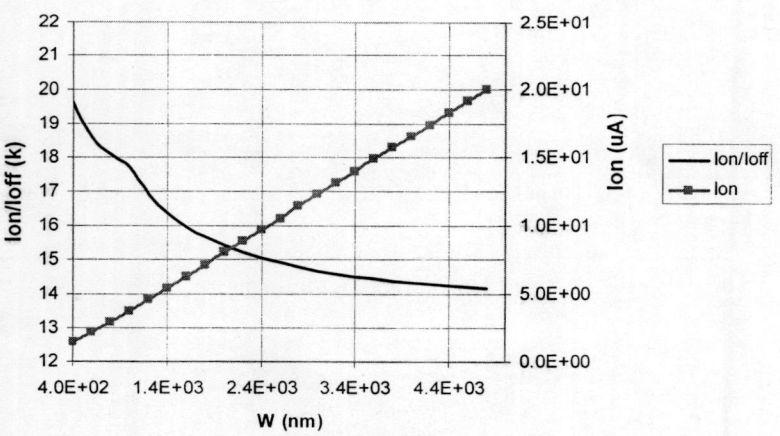

Figure 14-7   90nm pMOS Ion/Ioff and Ion vs. W

To produce the required drive current, a sleep transistor is commonly designed by connecting a number of small W transistors in parallel in a multi-finger style. The required drive current of a sleep transistor is defined based on statistical data of current consumption of those standard cells in the region that a sleep transistor drives. For example, if the sleep transistors are planned to be placed in every other row and at 50um pitch, the total current of the cells in 50um x 2rows area is the required drive current. Once the size (L*ΣW) of the sleep transistor is defined based on the required drive current, the area efficiency is determined by the layout implementation.

It is worth noting that Ion linearly increases with W as shown in Figure 14-7. Consequently, Ion/W becomes constant at given L and Vbb. This means that once L and Vbb are defined, the area efficiency is mainly determined by the layout implementation of the sleep transistor. These layout issues and their affect on area efficiency are described in more detail in Appendix "Sleep Transistor Design Methodologies."

It is important to understand that the sleep transistor area, namely the total size of all transistors, is mainly determined by the switch efficiency once the total driving current for a design is defined. Consequently, higher switch efficiency results in smaller area of the sleep transistors. An nMOS sleep transistor usually produces higher switch efficiency and hence smaller total transistor size than its pMOS counterpart.

14.2.3   Body Bias Considerations

Applying reverse body bias on the sleep transistor can increase the switch efficiency and reduce leakage significantly. The cost of reverse body bias in the header switch is significantly smaller than in the footer switch. This is because the N-well of the pMOS transistor is readily available for bias tapping in the standard CMOS process. As long as the N-well of the sleep transistor has enough space (based on the hot-well spacing rule) from the surrounding standard cells' N-wells, it can be tapped to its own body bias supply. On the other hand, the nMOS transistor does not have a well in the standard CMOS process. Consequently it becomes necessary to create wells for nMOS sleep transistors to allow separate body bias. This triple well CMOS process will result in higher chip fabrication cost and design complexity. It also introduces more process variations which affect design performance. As the result, pMOS header is preferable in reverse body bias applications.

14.2.4   System Level Design Consideration

In SoC designs, blocks usually communicate in the active-high interface protocols referencing common ground (VSS) as logic "0." In header switch implementations, all signal nets in power-gated blocks are settled at Vss which is convenient from a system design perspective.

The header switch also avoids potential signal integrity issues introduced by the virtual ground used in footer switch designs. Another advantage of using header switches is that it allows a simple design of a pull-down transistor to isolate power-gated blocks and clamp output signals at logic "0."

14.2.5   Recommendations – Header vs. Footer

- For those designs where area efficiency is the primary concern and reverse body bias is not available, the footer switch is a good choice.

- In other cases, particularly when system level design and IP integration are a primary concern, the header switch implementation is a good choice. The switch and area efficiency can be improved by applying reverse body bias.

- The header switch implementation is quite common in power-gating designs currently being implemented.

- It is worth mentioning that the choice of sleep transistors can be limited by the availability of low-leakage transistors in a given technology. Most technologies provide thick t_{ox} transistors for IO cells and thin t_{ox} process for core transistors. The thick t_{ox} transistor is much less leaky but also has less drive than the thin t_{ox} transistors. Implementing the sleep transistors with the thick t_{ox} transistors produces the lowest possible leakage. However, it requires a large sleep transistor area to deliver the required drive current. Consequently, high-V_T thin t_{ox} transistors are preferred in most power-gating designs to control area cost. The higher leakage in the thin t_{ox} transistor is mitigated by long gate and/or reverse body bias techniques.

- If the area efficiency is critical, W should be chosen as large as possible for a given cell height to form a single row of parallel transistors in the sleep transistor layout implementation.

- If the minimum standby leakage is the primary goal, then the optimal W (usually small) should be considered for high switch efficiency and hence low leakage. It is worth mentioning that both switch efficiency and leakage current becomes more sensitive to process variations with the reduction of W, particularly in sub-90nm region.

- For compromised area and leakage goals, the optimal W is obtained by investigating the area and leakage trade-offs of the sleep transistor with different W through SPICE analysis.

14.3 Rail vs. Strap VDD Supply

The sleep transistors get their power supply from the permanent power network (VDD) and deliver it to the virtual power network (VVdd) that drives the logic cells in a power domain. There can be two ways to distribute Vdd to the sleep transistors.

14.3.1 Parallel Rail VDD Distribution

In this implementation, a Vdd rail is added to a cell row in parallel with the VVdd rail. The sleep transistors get their permanent power supply by connecting to the Vdd rails as show in Figure 14-8.

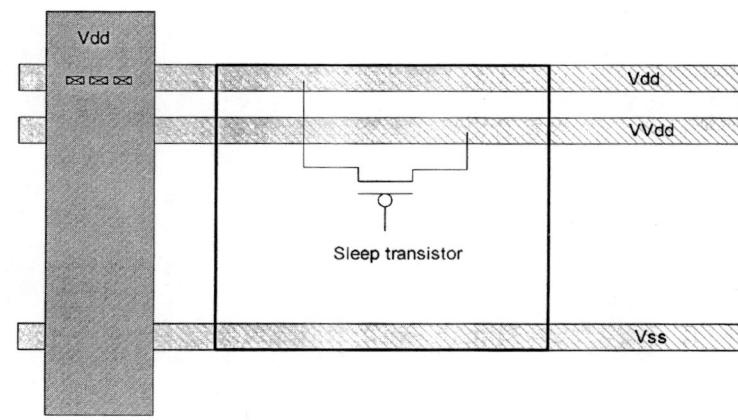

Figure 14-8   Parallel Vdd Rail Distribution

The Vdd network is built in the same way as a conventional power network from top metal layers down to the Vdd rail layer.

The advantage of this implementation is that the permanent power rail is reachable throughout the design. Consequently, the sleep transistors can be optimally placed without constraints based on accessing Vdd connections. It also enables the designer to move or insert a sleep transistor in post-layout to fix IR drop violations.

Moreover, there is no restriction on the placement of those special cells that require connections to the permanent power supply. This helps leverage conventional physical synthesis tools and flows.

However, this implementation takes at least one track of routing resources in every row in the Vdd rail layer.

In addition, it often incurs a layer conflict with conventional standard library cells which use the metal 1 layer for cell internal routing. In that case, it is necessary to create custom designed standard cell library where the added permanent power rail does not short other cell internal routes.

14.3.2 Power Strap VDD Distribution

In this implementation, a permanent power network is built in one or two top metal layers. The sleep transistors are placed under the straps of the coarse-grain network and get their Vdd supply through via pillars as illustrated in Figure 14-9.

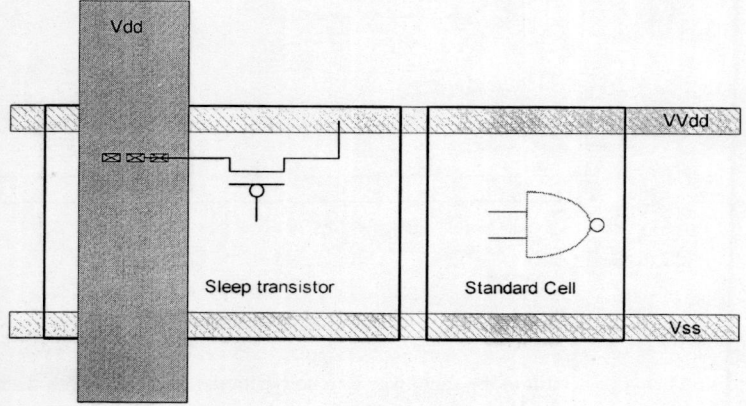

Figure 14-9   Power Strap Vdd Distribution

The virtual power network, driven by the sleep transistors, is built like a conventional power network from the sleep transistors' VVdd metal down to VVdd rails which connect to the standard logic cells. The sleep transistors are usually placed on a coarse grid; each drives the cells in a grid with a number of rows.

The main advantage of this implementation is that it allows the use of a normal standard cell library in a power-gating design. There is no need to add a second rail to the library cells. The library developer only needs to design a few switch cells, always-on repeaters and retention registers for the library.

However, the permanent power network no longer covers the design area. As a result, any cells that require a permanent power supply need either to be placed under the permanent power network or connected to it by power-routing. The former method imposes a placement constraint. The latter method introduces possible power integrity issues on the power-routing nets, which will require checking by IR drop analysis.

Moreover, the constraint that the sleep transistors must be under the permanent power network adds complexity in power network synthesis.

14.3.3 Recommendations for Supply Distribution

- If a special standard cell library which provides the extra Vdd rail is not available, the power strap Vdd distribution must be used.
- If the impact on routing resources becomes a concern, then the power strap Vdd distribution is often a good choice.
- If there are a significant number of retention registers in a design and power integrity in power-routing nets becomes a concern, the parallel rail distribution can ease the concern.
- It is worth mentioning that using small switch cells that can be placed in every row will simplify the virtual power network, which becomes a simple VVdd rail in every row.

14.4 A Sleep Transistor Example

An example of a double-row 90nm header switch cell is shown in Figure 14-10 [SALT]. In this design, 60 small pMOS transistors of 0.55um width form a sleep transistor as a 6-row transistor array. Normal body bias is applied to the sleep transistor, enabling the N-well to be extended around the sleep transistor to join adjacent tap-less standard cells and share well taps. In that case, Vss can be put in the middle of the two rows without causing significant area penalty. A pair of inverters that drive the sleep transistors is implemented in the cell for area efficiency.

The details of the sleep transistor design methods and guidelines are described in Appendix A.

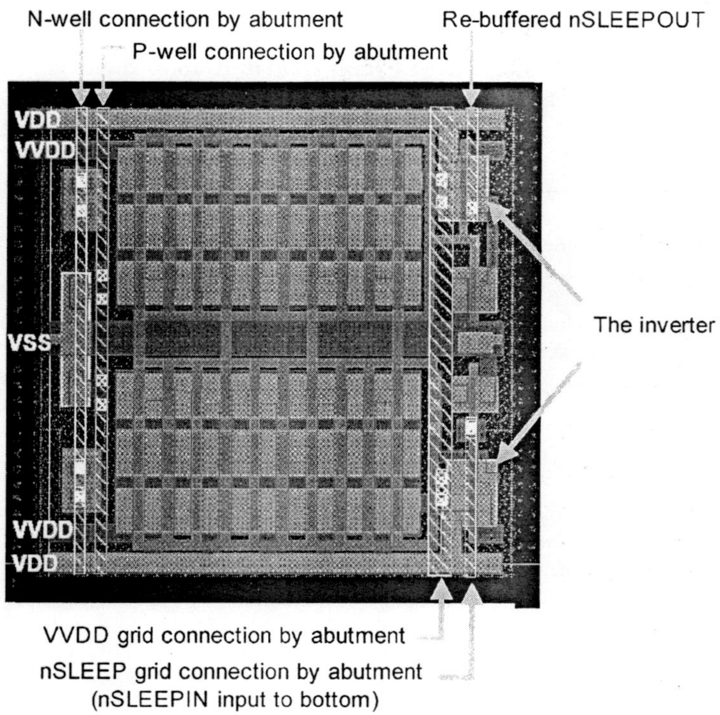

N-well connection by abutment   Re-buffered nSLEEPOUT
     P-well connection by abutment

VVDD grid connection by abutment
nSLEEP grid connection by abutment
(nSLEEPIN input to bottom)

Figure 14-10   A Double-row Sleep Transistor Implementation

14.5 Wakeup Current and Latency Control Methods

In a power-gating design, thousands of sleep transistors are commonly used to provide sufficient current to the design. When the design is coming out of sleep mode, the sleep transistors are switched on to supply power to the design. Simultaneously turning on the sleep transistors will result in a very large current (on the order of tens of amps) in charging the design to a full power-on state.

This large in-rush current will cause a large IR drop in the design, and can cause functional errors. In the worst case, the large current surge could result in short term VDD collapse, causing the state saved in retention registers and memories to be corrupted. Therefore, it is critical to limit in-rush current during power-on.

One possible way to control the in-rush current is to separate the chip power supply net into individual rows, each driven by a few sleep transistors. At power-up each row is turned on in sequence. This limits power-on current to the charge of one row at a time. However, this method has a major issue. Crowbar currents may occur in a powered-on row caused by floating inputs driven by other un-powered rows. These crowbar currents again can create an unacceptable IR drop. Consequently, this method has not been adopted in industrial power gating designs.

14.5.1 Single Daisy Chain Sleep Transistor Distribution

Another way to control in-rush current is to turn on the sleep transistors gradually as to prevent simultaneous switching current. This can be done by configuring the sleep transistors in a daisy chain style as shown in Figure 14-11.

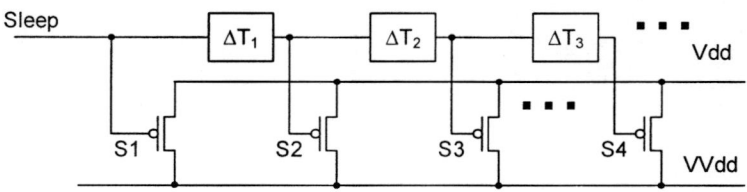

Figure 14-11   Single-daisy chain sleep transistor distribution

In this distribution, the sleep transistors are controlled by a daisy chain of delay elements and turned on one-by-one at interval of ΔT. Consequently, the charge current gradually increases with the number of turn-on sleep transistors. The delay elements are implemented by buffers at little extra cost because the buffers are needed to distribute the Sleep signal to every sleep transistors cross a chip.

This implementation is simple. However, the short delay of the buffers in the chain usually turns on the sleep transistors too quickly, resulting in larger than acceptable in-rush current at wakeup. To resolve the issue, the sleep transistors are commonly implemented in two daisy chains in industrial power-gating designs.

14.5.2 Dual Daisy Chain Sleep Transistor Distribution

The idea of the dual daisy chain distribution is to use weak transistors to trickle charge the design and hence prevent large in-rush current. Once the design is trickle charged close to Vdd, large transistors of the optimal drive strength are turned on to provide current for normal operation. In this approach, the sleep transistors are split into two chains: a weak transistor chain and main transistor chain as shown in Figure 14-12.

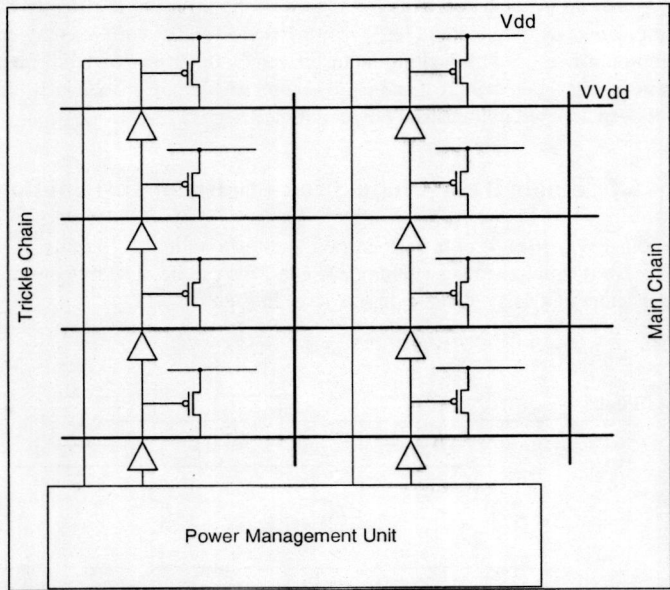

Figure 14-12   Dual Daisy Chain Sleep Transistor Distribution

The size of the weak trickle charge transistors is determined by the user-defined in-rush current limit and the maximum permissible turn-on delay time. A smaller trickle current may reduce the total surge current, but also increases the time taken to bring the system out of sleep mode. More details about in-rush current control methods are provided in the "Main chain turn-on control" section later.

The size of the sleep transistors in the main chain is optimized by the methods described in the previous section for the defined performance and leakage goals.

It is important to understand the differences of the sleep transistor designs in the trickle chain and main chain. The goal of trickle sleep transistor design is to control wakeup rush current and reduce wakeup latency due to trickle charge time. The trickle chain design involves transient simulation of design charge up in the wakeup period. More detailed discussions are given in the Appendix A. On the other hand, the main chain sleep transistor design is focused on meeting IR drop target and reducing the sleep transistor area. The sleep transistors are optimized in the active mode where all sleep transistors are turned on. In summary, the trickle and main sleep transistors

are designed for different goals and optimized in different operating modes or periods.

14.5.3   Parallel Short Chain Distribution of the Main Sleep Transistor

The wakeup latency is composed of the trickle charge time by the weak chain and the turn on time of the main chain. The size and number of the weak transistors are largely determined by the wakeup in-rush current limit and hence are constrained. As the result, the wakeup latency reduction is primarily determined by the distribution of the main chain.

The sleep transistors in the main chain can be configured as a single daisy chain; this approach takes the longest time to charge up the design but incurs the smallest peak charge current. On the other hand, the sleep transistors can be configured as a parallel array which simultaneously charges up the design; this approach has the smallest delay but the largest peak current.

The parallel short chain distribution of the sleep transistor in the main chain is a compromise of the two extreme cases. The sleep transistors are connected as a number of short daisy chains. These short chains are connected in parallel and turned on simultaneously when the main chain is turned on. As the result, the main chain power-on latency is reduced to the delay of a short chain. However, the peak current increases by the number of short chains as they are turned on simultaneously. The optimal number of the short chains can be obtained by SPICE analysis to limit the peak current to an acceptable level.

14.5.4   Main Chain Turn-on Control

Once the weak and main sleep transistors have been designed and the dual-daisy chain's distribution has been determined, we need to determine the threshold at which the main chain is turned on. A lower threshold causes the main chain to turn on earlier, but at the expense of higher the peak current. The optimal threshold is then determined by the peak current constraint.

14.5.5   Buffer Delay Based Main Chain Turn-on Control

A simple way to control the main header turn-on threshold is by controlling the time to trickle charge the design to the required threshold through the weak chain. In real power-gating designs, the trickle charge is controlled by the buffer chain which turns on the weak transistors in sequence. Consequently, the trickle charge time is determined by the buffer chain delay.

The drawback of the buffer delay based main chain turn-on control is that the buffer chain delay varies with the number of the sleep transistors in the chain, which is design dependent. Also, the delay is sensitive to PVT variations in the design. However, this style has least overhead cost, because the delay is a free: it is a product of the buffer chain, which is needed anyway as a part of the daisy chain.

14.5.6 Programmable Main chain Turn-on Control

One solution to this problem is to add a programmable delay element (typically a counter) to control the time to turn on the main chain. The counter can be programmed for the specific delay required to trickle charge the design to the desired threshold before turning on the main chain.

The main advantage of this method is its ability to define the optimal main chain turn-on time in different PVT conditions.

To reduce the counter's size, the counter can start counting after all the sleep transistors in the weak daisy chain are turned on, instead of at the beginning of wakeup. We can use the sleep signal at the end of the weak daisy chain as the start signal to the counter. This significantly reduces the counting period and hence the size of the counter.

The counter can be programmed by either software or hardware. The former is flexible and tunable for various designs and application environments. The latter has the advantage that no hardware-specific code needs to be added to the application software.

The program resolution of the main chain turn-on time is determined by the counter clock speed; the minimum delay delta is limited by the counter clock cycle time. If this clock does not provide adequate resolution, we can implement a fine-tune programmable delay element in addition to the counter. An example of such a fine-tune programmable delay element is a simple delay buffer chain with a mux to select different buffer segments and hence different delays.

14.5.7 Power-off Latency Reduction

Unlike power up, during power down we would like to switch off power quickly to eliminate leakage currents as soon as possible. This can be done by adding an OR gate to the sleep transistor control as illustrated by the diagram in Figure 14-13.

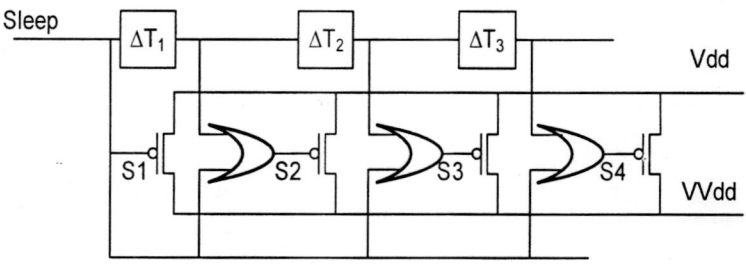

Figure 14-13   Parallel Switching off Sleep Transistors

However, adding an OR gate to every sleep transistor and routing them across the chip will incur a considerable area and routing resources penalty. A good compromise is to apply the method to each sleep transistor chain segment, that is, one OR gate for each short daisy chain of sleep transistors in a column, to simultaneously turn on all chain segments instead of all sleep transistors.

14.5.8 Recommendations for Power Switching Control

- For a large power-gating design, the dual daisy chain distribution is preferable to the single chain distribution since it provides much better in-rush current control.

- To determine the size of the weak sleep transistors, we recommend running SPICE on a small evaluation circuit to get an initial size and then verify the size by running SPICE on a good sized evaluation circuit. It is helpful to layout the evaluation design, push it down to transistor level view, extract RC on power and signal nets, and generate a SPICE deck with the RC annotations. For a large design, alternative analysis tools, such as NanoSim, which are able to simulate large design much more efficiently than SPICE, are recommended.

- The parallel short main chains often results in smaller wakeup latency while meeting the in-rush current constraint than the single main chain configuration. However, if the wakeup in-rush current constraint is tight, then the single chain main header distribution with turn-on threshold of 95% of VDD is a safe choice.

- The trickle charge weak switch cells should be configured as a number parallel daisy chains to evenly trickle charge the entire design and thus reduce possible crowbar current.

- The control signal buffers in the daisy chain should be implemented by a pair of inverters built inside the switch cell. Implementing the inverters with long gate length reduces leakage and increases propagation delay to help meet the timing target for the daisy chain.

- For main chain turn-on control, the programmable delay method is recommended for its ability to obtain design specific, optimal turn-on time that meets both the in-rush current constraint and the wakeup latency requirement.

- We recommend turning off a number of short chains rather than turning off all sleep transistors at once; this approach results in lower di/dt and less area and routing penalties.

- We recommend stopping clocks before switching off the sleep transistors, to let all switching activity die down. This minimizes the dynamic current in the design and consequent di/dt when the sleep transistors are turned off.

14.6 An Example of a Dual Daisy Chain Sleep Transistor Implementation

An example of the dual daisy chain sleep transistor implementation is the one used on the SALT chip, shown in Figure 14-14.

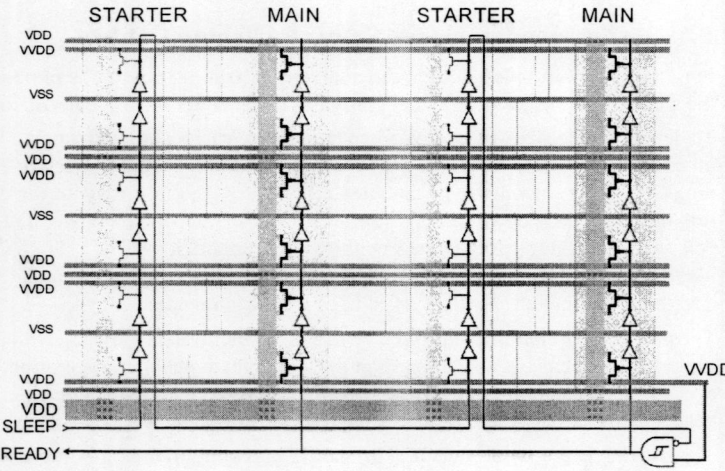

Figure 14-14   Dual Daisy Chain Sleep Transistor Implementation in SALT

In this example, the sleep transistors in the weak ("starter") chain and the main chain are placed in columns interleaving at a same pitch. The column chain segments of the weak sleep transistors are connected in series to form a long daisy chain for sequential trickle charging the design. However, the column chain segments of the main sleep transistors are connected in parallel. As a result, these column chains can be turned on

simultaneously to reduce time of charging VVDD from 90% to 100% VDD. The main sleep transistors in each column chain segment are configured as a daisy chain and switched on in sequence to reduce in-rush current.

APPENDIX A   *Sleep Transistor Design*

A sleep transistor is either a pMOS or nMOS high V_T transistor and is used as a switch to shut off power supplies to parts of a design in standby mode. The pMOS sleep transistor is used to switch VDD supply and hence is called a "header switch." The nMOS sleep transistor controls VSS supply and hence is called a "footer switch." In designs at 90nm and below, either a header or footer switch is used due to tight voltage margin and too large area penalty when both header and footer switches are implemented.

Although the concept of the sleep transistor is straight forward, optimal sleep transistor design and implementation are a challenge due to various effects, introduced by the sleep transistor and its implementations, on design performance, area, routability, overall power dissipation, and signal/power integrity.

Optimal sleep transistor design also depends on design specific goals and chosen CMOS technology and process. A number of decisions need to be made including the choice of header or footer switch, normal or reverse body bias, optimal transistor size, and layout implementation details such as single or double row and extra rail or direct via-pillar for permanent power connection.

This chapter is written as a guide to the advantages and trade-offs associated with different choices, rather than the design procedure which is well understood by the transistor designers. The investigations of various sleep transistor characteristics in the power-gating design context are based on SPICE analysis instead of the theories and equations. Device modeling becomes so complex in sub-90nm technology that the process parameter based device models are the best analysis tool for a quality industrial design.

A.1 Sleep Transistor Design Metrics

Quality of a sleep transistor design is often measured in terms of three metrics: switch efficiency, area efficiency and IR drop. The sleep transistor is optimized in gate length, width, finger size and body-bias to achieve high switch and area efficiencies, and low leakage current and IR drop.

A.1.1 Switch Efficiency

The sleep transistor efficiency (switch efficiency) is defined by a ratio of drain current in ON and OFF states, i.e. Ion/Ioff. It is desirable to maximize the efficiency to achieve high drive in normal operation and low leakage in sleep mode. The sleep transistor efficiency varies with gate length (L), width (W) and body bias (Vbb). The optimal values of L, W and Vbb vary with the technology and process. These optimal values are obtained by switch efficiency analysis in SPICE by sweeping L, W and Vbb and measuring Ion/Ioff in each case.

An example of Ion/Ioff SPICE analysis circuit is shown in Figure A-1.

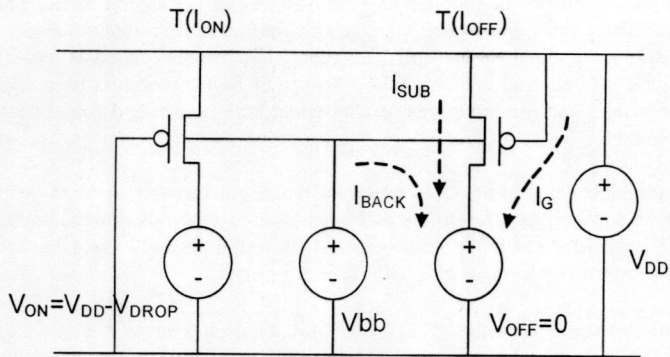

Figure A-1   An Example of Ion/Ioff SPICE Analysis Circuit

The circuit is composed of two high V_T transistors which are configured in ON and OFF state respectively. The bodies of both transistors are biased at Vbb which is swept in SPICE analysis. The ON-state transistor T(Ion) is configured for Ion analysis. A voltage source Von is added to bias Vds at a specified IR drop target, e.g. 10mV. It also used to measure drain current as Ion. The gate leakage current is not included in Ion, because it does not contribute to drive current for logic operations in a power-gating design. A high temperature of 125°C is set on the ON-state transistor to model high chip temperature in operating mode.

The OFF state transistor is configured for Ioff analysis. In this case, a zero voltage source Voff is inserted to measure the drain current as Ioff. Since Source, Gate and Body of the transistor are all biased at or above Vdd in the OFF configuration, the drain current collects all main leakage currents including the gate tunneling current Ig, sub-threshold channel current, Gate-Induced-Drain-Leakage (GIDL) current and drain-substrate Band-To-Band-Tunneling (BTBT) current. A room temperature of 25°C is set on OFF state transistor to reflect the cool situation when the design is in sleep mode.

The gate of the transistors could be driven by inverters in logic "1" and "0" states reflecting the real driving situation in a power-gating design. However, this is not necessary. The very small Vds (about 10mV) of the inverters has little impact on the Vgs of the ON and OFF transistors and hence negligible errors in the simulation results.

An example of a switch efficiency curve is shown in Figure A-2. This curve shows a 65nm high V_T pMOS transistor in a nominal or generic process. The Vbb was biased at Vdd corresponding to the normal body bias condition in SoC designs. The gate length and width were swept from 65nm to 165nm and 0.4um to 5um respectively.

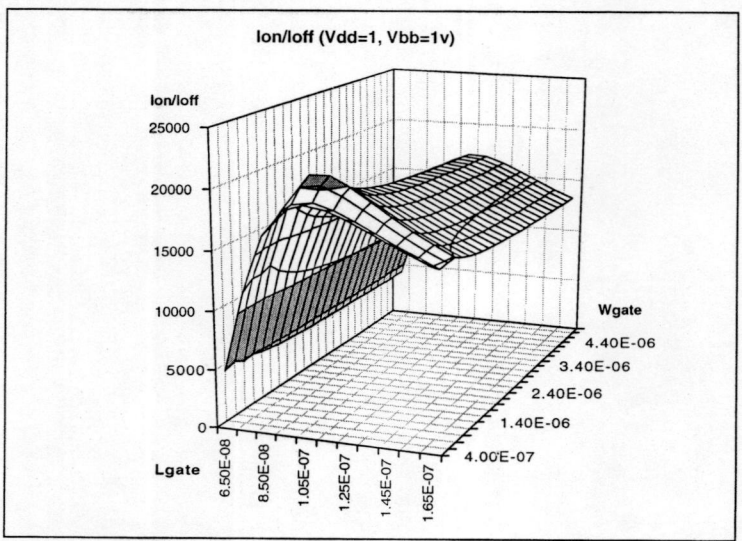

Figure A-2   65nm High V_T pMOS Ion/Ioff Curve (normal back bias)

The switch efficiency increases with L and reaches peak value when L= 105nm. This is mainly due to consequent V_T increase with L which results in exponentially reduction of sub-threshold leakage current and linear reduction of Ion (in linear region).

However, the efficiency declines when L is longer than 105nm. This is because L enters the region where Vth increase is saturated but channel resistance continues to increases. As the result, Ioff reduction becomes less significant than the drop of Ion.

The efficiency also depends on gate width. The Ion/Ioff curve in Figure A-2 on page 251 shows that Ion/Ioff is relatively higher in narrow gate widths of up to 1.4um and peak at W=1um. This is mainly due to the fringing effect of narrow gate width on V_T. The narrow width effect becomes complex in sub-65nm which results in the unusual Ion/Ioff behavior in sub-micro narrow gate width transistors. Wider than 1.8um, Ion/Ioff becomes not sensitive to the gate width.

The switch efficiency also depends on body bias. Reversed body bias (RBB) increases V_T and hence reduces the sub-threshold leakage resulting in higher switch efficiency. The effect of RBB varies significantly with technology and process, specially in sub-90nm nodes. It is important to evaluate the effect of body bias on the switch efficiency in the chosen technology and process by SPICE analysis of the circuit in Figure A-1. with various body bias voltages. An example of such an analysis result on a 65nm high V_T pMOS transistor at 1.4V body bias is shown in Figure A-3.

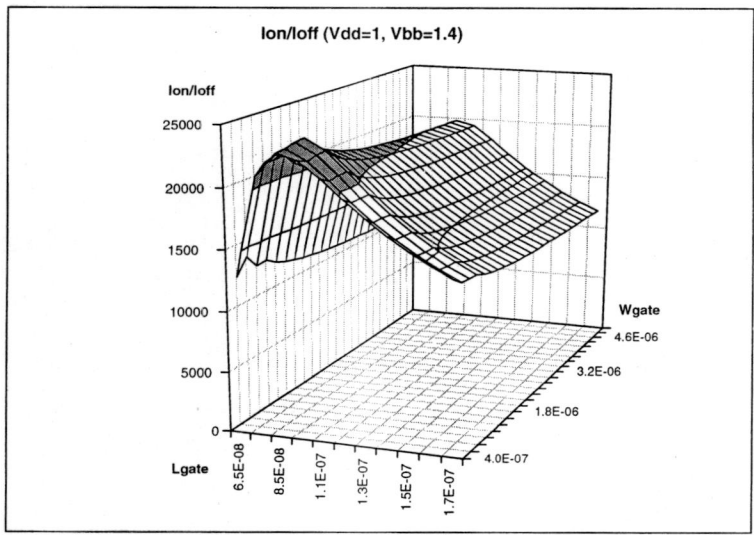

Figure A-3   65nm pMOS Ion/Ioff Curve at Reverse Back Bias (Vbb=1.4V)

The result shows that Ion/Ioff at 1.4V body bias increases by 16% compared with max Ion/Ioff in the normal body bias case. Moreover, the max Ion/Ioff occurs at a smaller gate width of 85nm compared with 105nm in the normal body bias case. Con-

sequently, the sleep transistor of same drive current is smaller and more efficient with reversed body bias. However, increase body bias beyond an optimal value will result in Ion/Ioff reduction. This is because the V_T reduction becomes less significant at high RBB, which results in slower reduction of the sub-threshold leakage current. On the other hand, BTBT leakage current increases with the RBB and becomes dominant at high RBB due to the high electric field generated by RBB at brain-substrate junction.

A.1.2   Area Efficiency

The area efficiency of a sleep transistor is defined by the ratio of its drive current and silicon area, i.e. Ion/Asleep. The Ion is the drain current when the transistor is in ON state and is biased at Vds equaling IR drop target. Asleep is the silicon area of the sleep transistor depending on the gate length (L), width (W) and layout implementation. To produce required drive current, a sleep transistor is commonly designed by connecting a number of small transistors in parallel in a multi-finger style. Certain area overhead caused by design rules, such as hot-well spacing and diffusion spacing rules, applies to the layout implementation of the sleep transistor and affects the area efficiency. The detailed discussions on design for area efficiency will be provided later in this chapter.

A.1.3   IR Drop

The IR drop on the sleep transistor is mainly determined by the equivalent channel resistance (Ron = Vds/Ion) when the sleep transistor is conducting. The smaller of Ron, the smaller of IR drop. Gate width is the primary parameter to determine Ron. However, gate length also affects Ids and hence Ron. In a sub-50mV Vds region corresponding to ON state of the sleep transistor, Ron linearly increases with gate length and body bias as shown by solid lines in Figure A-4 on page 254. This is the result for a 2.2um wide pMOS high V_T transistor in a representative process.

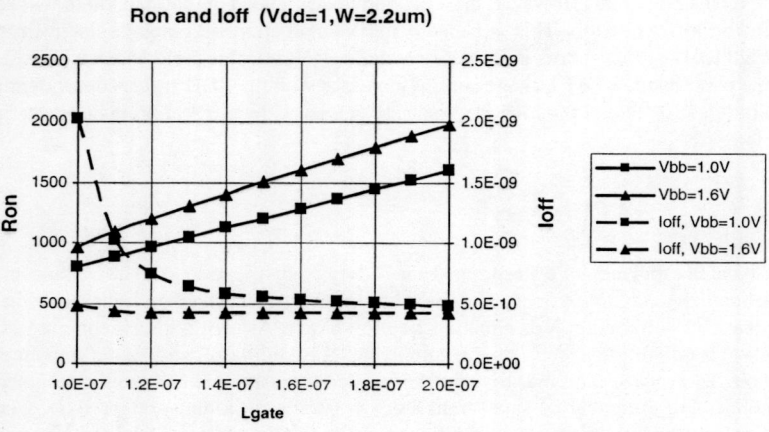

Figure A-4   90nm High V_T pMOS Ron and Ioff Curves

The result shows that Ron is more sensitive to gate length than body bias. From the Ron and leakage curves in Figure A-4, we can see that at a same leakage current of 0.5nA (triangle and square marked dash lines), Ron is 1K Ohm (triangle marked solid line) in the sleep transistor of 100nm Lgate and 1.6V body bias compared with 1.5K Ohm (square marked solid line) in the sleep transistor of 180nm Lgate and normal (1V) body bias. This shows that Ron is usually smaller in short gate at reverse body bias than in the long gate at normal body bias.

A.1.4   Normal vs. Reverse Body Bias

The substrate of the sleep transistor can be biased normally by tapping the substrate or well to the permanent Vdd rails in pMOS sleep transistor and permanent Vss rails in nMOS transistor. It can also be reverse biased by connecting to a separate bias voltage higher than Vdd in pMOS transistor and lower than Vss in nMOS transistor.

A transistor's leakage current reduces with the reverse body bias, as shown by the Ioff/Vbb curve in Figure A-6. The leakage reduction is mainly due to the increase of Vth with the reserve body bias.

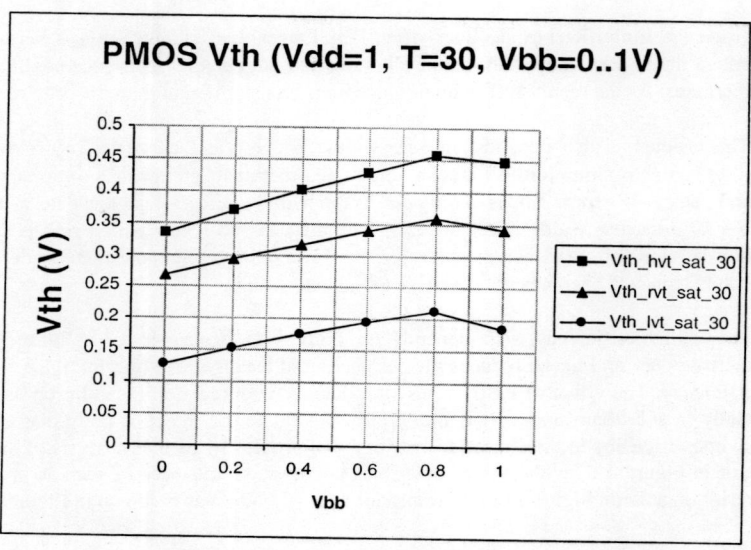

Figure A-5

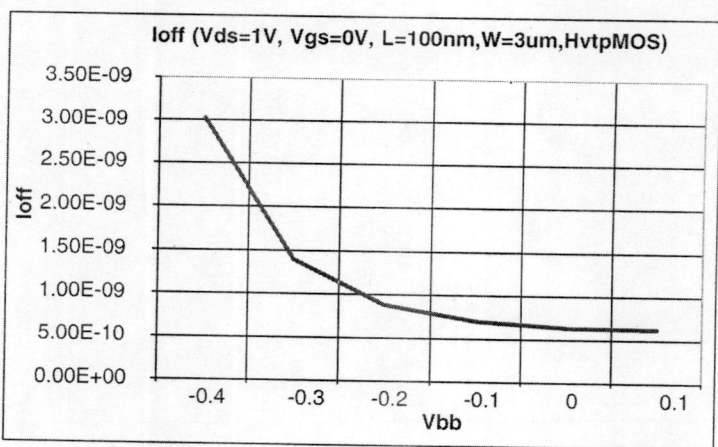

Figure A-6   90nm Ioff/Vbb Curve of High-V_T pMOS (L=100nm, W=3um, Vds=1V, Vgs=0V, tt-corner)

The main advantage of normal body bias is simplicity in implementation. No special bias voltage sources nor bias supply distribution is required. On the other hand, the reverse body bias usually results in higher switch efficiency, lower leakage and smaller area in tap-cell implementations. The choice of normal or reverse body bias depends on design goals and trade-offs which are discussed below.

Ioff decreases exponentially with the increase of V_T while Ion reduces linearly with the Vth increase. Therefore, we can find an optimal V_T for a given technology and process to obtain maximum switch efficiency (Ion/Ioff). V_T increases with gate length (L) and reverse body bias (Vbb). The optimal Vth can be produced from the optimal combination of L and Vbb. The best way to find the optimal Vbb and L is to run SPICE simulation and analyze Ion/Ioff with different Vbb and L. Figure A-6 and Figure A-7 show the switch efficiency and Ioff curves for a high Vth pMOS transistor in a 90nm process and a 65nm process at typical process corner respectively. The transistor is 2.2um wide with Vds biased at 1.0V.

In a typical process (Figure A-7), the maximum Ion/Ioff increase with Vbb and reaches peak value of ~22,000 at Vbb = 1.8V and gate length L=100nm. This is about 43% improvement compared with the max Ion/Ioff of ~15,000 in the normal body bias case of L=140nm. The Ioff is also reduced by 23%. In this case, the short channel sleep transistor at reverse body bias is a better choice than the long channel transistor at normal body bias.

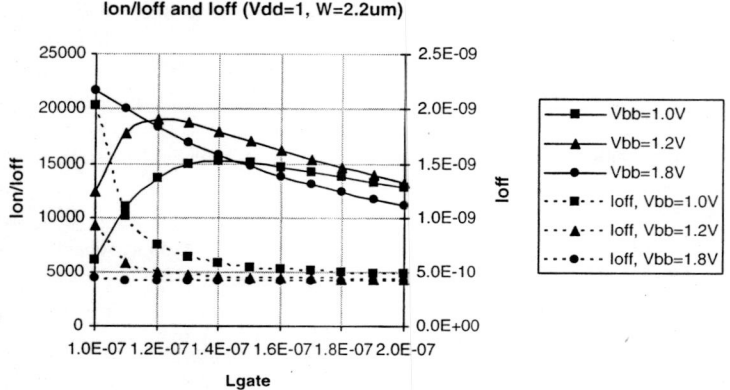

Figure A-7   90nm pMOS Ion/Ioff and Ioff Curves

However, the advantage of applying reverse body bias becomes significantly smaller in a 65nm pMOS transistor as shown in Figure A-8. Comparing the max Ion/Ioff of ~21000 in the reverse body bias case (Vbb=1.4V and L=85nm) and ~18000 in normal body bias case (Vbb=1.0V and L=105nm), the switch efficiency improvement is dropped to 15%. Ioff reduction ratio is also reduced to 9.8%. The main reason for the reduction of the effect is the significant increase of BTBT leakage currents in 65nm technology where strong halo doping technique is commonly used to reduce short channel effect and subthreshold leakage current at the price of BTBT increase. Moreover, t_{ox} reduction with technology scaling causes higher GIDL which does not decline with RBB. Consequently, BTBT and GIDL leakage current are becoming dominate leakage at RBB and applying RBB would be less and less effective to improve Ion/Ioff in sub-45nm technologies.

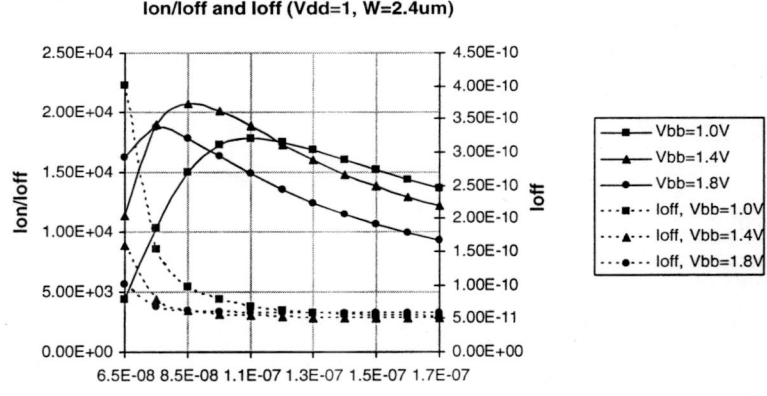

Figure A-8   65nm pMOS Ion/Ioff and Ioff Curves

It is worth mentioning that nMOS transistor behaves differently. Figure A-9 shows the nMOS switch efficiency curve for a 65nm process. Comparing the Ion/Ioff of the reverse body biased transistor of 85nm gate length with the normal body biased transistor of 105nm gate length, it shows that the reverse body bias still produce 34% higher Ion/Ioff than the normal body bias. Ioff is also improved by 33%. With technology scaling below 65nm, more and more complex processes, such as NMOS/PMOS channel engineering and halo doping, have been applied to the devices to address critical issues in areas such as short channel effect, leakage control and power-delay product improvement. Such complex processes significantly change transistor's behavior and complicate device modeling. Therefore, it is essential to run

SPICE analysis on the design specific devices to obtain realistic transistor characteristics.

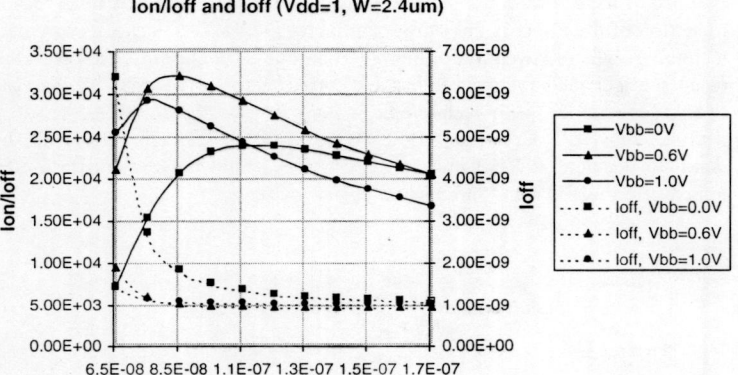

Ion/Ioff and Ioff (Vdd=1, W=2.4um)

Legend:
- Vbb=0V
- Vbb=0.6V
- Vbb=1.0V
- Ioff, Vbb=0.0V
- Ioff, Vbb=0.6V
- Ioff, Vbb=1.0V

Figure A-9   65nm nMOS Ion/Ioff and Ioff Curves

The advantages of the reverse body bias sleep transistor include smaller size than the normal bias sleep transistor for its higher Ion/(L*W) ratio.

Moreover, the reverse body bias sleep transistor has smaller Ron and hence IR drop than the normal body bias counter part at a same leakage current.

Another advantage of the reverse body bias technique is that it becomes applicable to tune body bias voltage for optimal leakage and performance in different modes, such as applying the reverse body bias in standby mode to minimize leakage current and applying normal or forward body bias to produce maximum drive current in normal operation mode. This adaptive body bias method further increases Ion/Ioff significantly.

However, the reverse body bias method introduces significant trade-offs below:

Firstly, it requires separate Vbb voltage source which is out of the range of the power supply voltage to the design, e.g., Vbb > Vdd in header switch implementations.

Secondly, it requires a dedicate Vbb distributions to every sleep transistor in the design. The Vbb network needs to be planned carefully to prevent from noise injec-

tions to it, because the noise in Vbb can cause significant Vth and cell performance variations.

Thirdly, it requires costly triple well process in the footer switch implementation to create separate P-well in the nMOS sleep transistors. This is not a problem in the header switch.

Finally, it imposes hot-well spacing rule between the sleep transistor and adjacent standard cells. This boundary are overhead becomes considerable in small sleep transistor designs.

On the other hand, the normal body bias application does not require body bias voltage source nor the bias distribution. The well is simply tapped to permanent Vdd. In the tap-less standard cell designs, the sleep transistors and standard cells can share well which eliminates the well-spacing constraint. The sleep transistors' well is tapped at intervals with standard cell's well.

A.1.5   Recommendations

- We recommend running SPICE on the high-V_T transistors in the chosen library to analyze the body bias effect which varies considerably in different technologies, process and foundries.

- If minimum leakage is the primary concern, then the reverse body bias is the correct choice.

- In 90nm technology, the reverse bias is usually a good choice for its lower leakage and area overhead than the normal bias. The advantages become considerably smaller in 65nm technology where long gate sleep transistor at normal body bias might be preferable than the reverse body biased choice. The decision should be made on overall considerations of design specific goals and trade-offs in normal and reverse body bias designs.

- If a separate voltage source, either on-chip or off-chip, is difficult to obtained, then the normal body bias should be chosen.

- If the footer switch has been chosen for the power-gating design, the normal body bias should be used, unless the standard cell library support separate P-well connection in the nMOS sleep transistor for applying reverse body bias.

- If tap-less standard cell libraries are used, the normal body bias sleep transistor becomes considerable for its higher area efficiency than the reversed body bias sleep transistor.

- In the reversed body biased sleep transistor applications, the body of the sleep transistor, such as N-well of a pMOS) must be separated with good space (Hot-well spacing rule) from the bodies of the standard cells next to it to prevent possible latch-up events due to difference body bias potentials at the sleep transistors and the standard cells.

- In the normal body biased sleep transistor applications where tap-less standard cell libraries are used, the body of the pMOS and nMOS transistors in the standard cells must be tapped to the permanent VDD and VSS respectively to maintain safe bias voltage as the sleep transistors when the design is in the sleep mode. On the other hand, for those applications which use those standard cell libraries where the body has been connected to the source of the transistors in the standard cells, the body of the sleep transistor must be separated from the standard cells in a same way as for the reversed body biased sleep transistor design. This is because the bodies of the standard cells will be floating to VDD in a footer gating design and VSS in a header gating design in the sleep mode and hence creating large bias potential difference between the sleep transistor and the standard cells.

A.2   Layout Design for Area Efficiency

Area efficiency depends on the size (L*W) and layout implementation of the sleep transistors. The optimal L is determined by the switch efficiency and can be obtained from the switch efficiency curves generated from SPICE analysis. Once L is defined, the area efficiency is mainly determined by the transistor width W and layout implementation. The method for optimal sleep transistor size (L*ΣW) for high area efficiency has been described in Chapter 4 and hence is not repeated here. Once the size (L*ΣW) of the sleep transistor is determined, the area efficiency of the sleep transistor is dependent on layout implementations described below.

Two design rules affect the area efficiency in the layout implementation. The first is the active-area spacing rule which defines the minimum space between diffusion regions of two transistors. This introduces area overhead in space between rows of the parallel small W transistors in the layout implementations. The second rule is the hot-well spacing rule which defines the minimum distance from N-well in the reverse body biased sleep transistor to the N-wells in the adjacent standard cells, due to different well bias voltages in the sleep transistor and the standard cells. The spacing rule introduced area overhead in a sleep transistor is shown in Figure A-10.

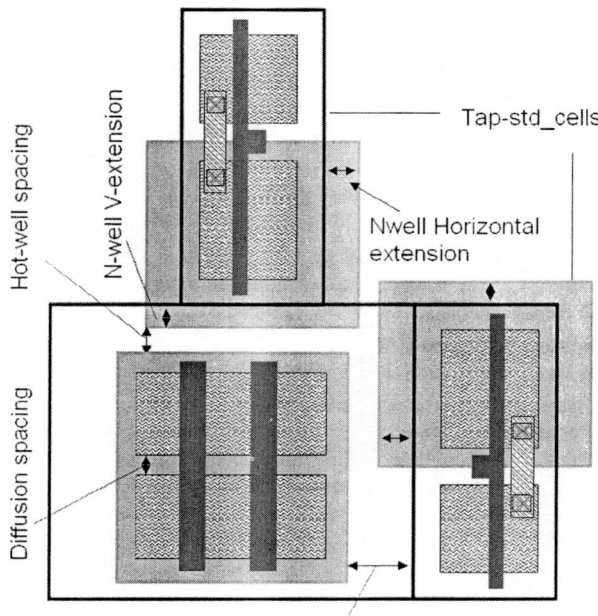

Figure A-10   Spacing rule overhead of sleep transistor with well-tapped standard cells

To reduce effect of the spacing overhead on the area efficiency, we can increase the size of the sleep transistor so as to reduce the ratio of the spacing overhead over the sleep transistor. This means that the horizontal size of the sleep transistor needs to be large enough to mitigate the effect of the spacing overhead at both sides of the cell and W should be large enough to avoid multiple rows so as to eliminate the diffusion spacing overhead. If the standard cell library is designed without individual well tap in a cell (tap-less cells), then the well spacing rule becomes not applicable to the normal body biased sleep transistor because the sleep transistor and standard cell share well taps and are biased at same voltage (permanent Vdd). Consequently, the boundary are overhead is only caused by the diffusion spacing rule.

It is worth noting that Ion linearly increases with W as shown by the dashed line in Figure A-8 on page 257. Consequently, Ion/W becomes constant at given L and Vbb which means that the area efficiency is mainly determined by the layout implementation of the sleep transistor once L and Vbb are defined.

A.2.1   Recommendations

- If area efficiency is critical, W should be chosen as large as possible for a given cell height to form a single row of parallel transistors in the sleep transistor layout implementation.

- If minimum standby leakage is the primary goal, then the optimal W (usually small) should be considered for high switch efficiency and hence low leakage. It is worth mentioning that both switch efficiency and leakage current becomes more sensitive to process variations with the deduction of W particularly in sub-800nm region.

- For compromised area and leakage goals, the optimal W is obtainable by investigating the area and leakage trade-offs of the sleep transistor with different W through SPICE analysis.

- A larger sleep transistor ($L*\Sigma W$) is more area efficient because the spacing overhead has less impact on area efficiency.

- If the sleep transistors are implemented as a daisy chain (described in detail later) in the power network, the repeater that controls the sleep transistor is preferable to be implemented together with the sleep transistor as a single switch cell for area efficiency, because the space between the sleep transistor and the repeater need not follow the hot-well spacing rule.

- The sleep transistors are often implemented in two daisy chains which will be fully explained in the sleep transistor implementation section in the chapter. At the wakeup, a weak sleep transistor chain is turned on to trickle charge the virtual power network until 95% Vdd when a main chain is turned on to fully charge the network for operation. In this case, we recommend implementing the weak and main sleep transistors in the same switch cell to avoid the extra boundary spacing overhead which incurs the two transistors are separately implemented.

A.3   Single Row vs. Double Row

For high area efficiency, the sleep transistor can be designed twice as high as a standard cell. The double-row implementation has two advantages. Firstly, it eliminates the space imposed by the well spacing rule between the two rows because both rows are occupied by the same well of the sleep transistor. This is particularly effective in the reverse body bias implementation where the "hot-well" spacing is significant.

Secondly, W can extend into two row height forming a wide single row transistor array and hence high area efficiency.

The trade-off of the double-row implementation is the requirement of cell alignment with standard power and ground rails. This is usually not difficult to handle by the back-end flow.

The choice of single and double-row implementation also depends on how the sleep transistors are implemented in the power network of the design. Fewer of larger sleep transistors placed in coarse power network grids is usually more area efficient than more of smaller sleep transistors placed in fine grids, because the spacing overhead becomes less significant in a larger transistor. The maximum sleep transistor size and placement grid are constrained by impact on routability and IR drop at center of a power grid that a local sleep transistor drives.

A.3.1   Recommendations

- If the reverse body biased sleep transistor is chosen, the double-row implementation is a right choice. The Vdd should be at the middle of the two rows to allow W to be extend into two rows.

- If the sleep transistor is normal biased and placed at small power grid in the tapless standard cell design, the single-row layout is a good choice.

- However, if the tap-cell standard cell libraries are used, then double-row implementation usually produce higher area efficiency even with the normal body biased sleep transistor. This is because the N-well of the tap-cells is tied to the virtual power network while N-well of the sleep transistor is connected to the permanent power network. In the sleep mode, wells of the standard cells and the sleep transistors are biased at different voltages. Consequently, the hot-well spacing rule is imposed to the space between the sleep transistor and adjacent standard cells in the same way in the reverse biased sleep transistor applications.

A.4   In-rush Current and Latency Analysis

A practical way to obtain a desirable size of the weak sleep transistor in a trickle chain and the trickle chain configuration is to run SPICE transient simulation to evaluate the charge current and time of a block of cells driven by a trickle-charge transistor of various gate widths in a defined trickle charge chain. Then, we can generate rush current and charge-up time curves and choose the right gate width which meets both in-rush and wakeup latency requirements, from the curves. An example of such an evaluation circuit is shown in Figure A-11.

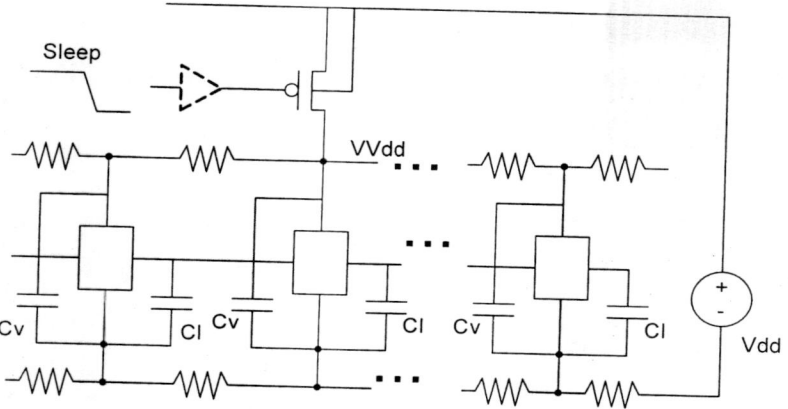

Figure A-11   Charge Current and Time SPICE Transient Simulation Circuit

The logic cells (shown as boxes in Figure A-11) are the collection of various types of standard cells. To simulate the transient behavior of the sleep transistor and the circuits of the logic cells during wakeup period, the transistor level models of the logic cells should be used in the evaluation circuit. The number of the cells is determined by the region of a sleep transistor network grid (such as 2 rows and 1 column of the power-grid) that the sleep transistor drives. The power network resistance can be obtained from rail extraction of an existing design with similar power rails. Cv models the power capacitance including de-cap. Cl is the load capacitance a cell drives. The charging time can be obtained by measuring VVdd voltage changes from 0 to 99% Vdd. It is worth mentioning that decap often dominates the power network capacitance in industrial designs. Consequently, the wakeup time and in-rush current vary significantly with decap. It is recommended have good estimations of decaps and include them in the wakeup in-rush current simulation so that the simulation results correlate with final design.

The single-grid based sleep transistor evaluation model is small and efficient for initial evaluation. However, it does not accurately represent the full scale power-gating where the sleep transistors are turned on in sequence and each turn-on transistor drives the whole power network. Therefore, a larger scale model of closing to the full design is often built for a final power-gating evaluation, once an initial evaluation has done. The larger model can be built by configuring the single-grid model into a sleep transistor network with buffer chains driving the sleep transistors in a defined sequence. The voltage and current curves in Figure A-12 and Figure A-13 are the result of SPICE simulation of such a circuit which has logic cells in 6 rows by 40 columns driven by 12 sleep transistors configured each driving logic cells in 2 rows by

10 columns. The results show the smaller transistors have lower current peak, but take much longer to fully charge the design. The optimal transistor size can be obtained from the W swept results based on the in-rush current constraint.

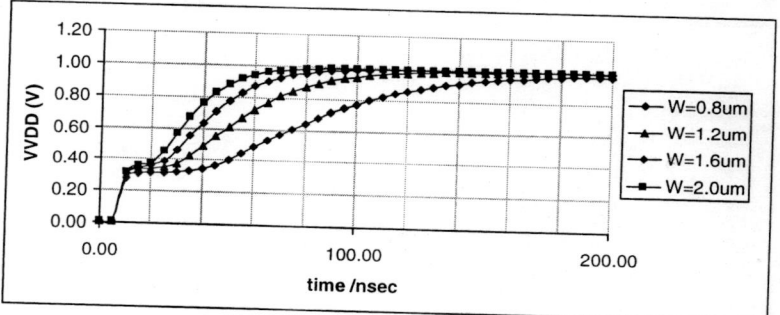

Figure A-12   Wakeup Power-on Voltage Curve

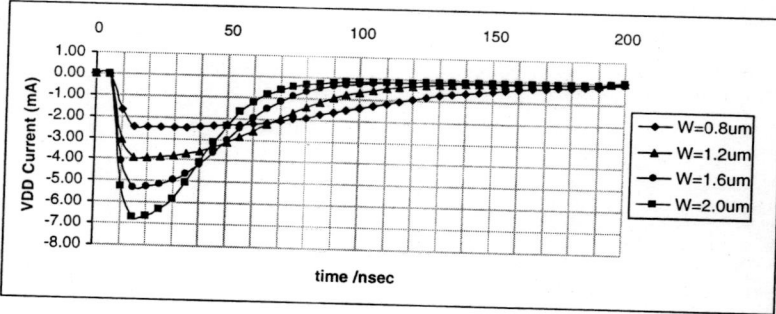

Figure A-13   Wakeup Power-on Current Curve

The shortcoming of the small SPICE evaluation of wakeup power-on sequence response is that it might not reflect exactly the power-on behavior of the chip design which is normally much larger. Fortunately, advanced dynamic power network analysis tools are capable of performing transient analysis of a large design. Consequently, it becomes practical to perform wakeup power-on sequence analysis on an actual design. In that case, the power network, the sleep transistors and logic cells can all be extracted from the layout with parasitics for accurate wakeup rush current and charge-up time analysis.

UPF Command Syntax

This chapter describes the syntax for selected UPF commands referenced in the text.

Excerpts from the "Unified Power Format(UPF) Standard, Version 1.0" reproduced by permission of Accellera. Copyright(c) 2006-2007 by Accellera. Accellera does not warrant or represent the accuracy or content of the excerpted material, and expressly disclaims any express or implied warranty. Accellera Standards excerpts are supplied "AS IS."

The full standards document in its entirety can be found under www.accellera.org.

B.1   add_pst_state

Purpose	Define the states of each of the supply nets for one possible state of the design	

Syntax	**add_pst_state** *state_name*
	  -pst *table_name*
	  -state *supply_states*	

Arguments	*state_name*	The power state.
	-pst *table_name*	The power state table (PST) to which this state applies.
	-state *supply_states*	The list of supply net state names , listed in the corresponding order of the **-supplies** listing in the **create_pst** command.
Return value	Return a 1 if successful or a 0 if not.	

The **add_pst_state** command defines the states of each of the supply nets for one possible state of the design.

It is an error if the number of *supply_state_names* is different than the number of supply nets within the PST.

Syntax example:

```
create_pst            pt -supplies { PN1   PN2   SOC/OTC/PN3 }
add_pst_state s1 -pst pt -state   { s08   s08   s08         }
add_pst_state s2 -pst pt -state   { s08   s08   off         }
add_pst_state s3 -pst pt -state   { s08   s09   off         }
```

B.2   connect_supply_net

Purpose	Connect a supply net to supply ports and/or pins	

Syntax	**connect_supply_net** *net_name*
	  [**-ports** *list*] [**-pins** *list*]
	  [<**-cells** *list* |**-domain** *domain_name*>]
	  [<**-rail_connection** *rail_type* | **-pg_type** *pg_type*>]*
	  [**-vct** *vct_name*]	

Arguments	*net_name*	The name of the supply net.
	-ports *list*	A list of ports to connect.
	-pins *list*	A list of pins on cells to connect.
	-cells *list*	A list of cells to use for **-rail_connection** or **-pg_type**.
	-domain *domain_name*	The domain to use for **-rail_connection** or **-pg_type**.
	-rail_connection *rail_type*	The rail type (for older libraries).
	-pg_type *pg_type*	The power/ground pin type.
	-vct *vct_name*	A VCT defining how values are mapped from UPF to an HDL model or from the HDL model to UPF.
Return value	Return the fully qualified name of the supply net if successful or a 0 if not.	

Any ports, pins, cells, supply nets, or domains are referenced relative to the current UPF scope .

The **connect_supply_net** command provides an explicit connection of a supply net to any port and overrides (has higher precedence than) the auto-connection semantics that might otherwise apply. **–domain** or **-cells** is required when the **-rail_connection** or **-pg_type** options are specified.

Use:

> **-ports** to connect to supply ports;

> **-pins** to connect to pins on library cells;

> **-cells** to connect to all pins of the appropriate type (power or ground) on the specified cells;

–rail_connection to connect to pins having this (rail) type; only use this if the **-cells** or **-domain** options are specified;

–pg_type to connect only to pins having this power/ground type (the pin type defined in the library model is used for determining the connection); only use this if the **-cells** or **-domain** options are specified;

-vct to indicate for every HDL port to which the port is connected, the supply net state shall be converted if it is being propagated into the HDL port, or the HDL port value shall be converted if it is being propagated onto the supply net. **-vct** is ignored for any connections of the supply net to pins or supply ports defined in UPF.

The following also apply.

— The **–ports** and **–pins** options are mutually exclusive with the **–cells**, **–domain**, **-rail_connection**, and **-pg_type** options.

— The **-rail_connection** and **-pg_type** options are mutually exclusive with each other.

— A supply net can only be connected to a port, pin, or cell in the same *extent* as the domain of that supply net.

— It is an error if *net_name* has not been previously created; in this case, a 0 shall be returned.

— It is an error if any design element specified in this command does not exist.

— It is an error if the value conversions specified in the VCT do not match the type of the HDL port.

Syntax examples:

```
connect_supply_net v09 -ports {VDD U18/v9 U21/v9}

connect_supply_net pd1_vdd
  -ports pll_inst/vdd
  -vct upf2vlog_vdd
```

B.3   create_power_domain

Purpose	Define a power supply distribution network for a set of design elements	

Syntax	**create_power_domain** *domain_name*
	    [-elements *list*]
	    [-include_scope]
	    [-scope *instance_name*]	

Arguments	*domain_name*	The new power domain; this shall be a simple (non-hierarchical) name.
	-elements *list*	Use this set of design elements in the power domain.
	-include_scope	Include the scope of the domain in the extent of the power domain.
	-scope *instance_name*	Create the power domain within this logic hierarchy scope.
Return value	Return the fully qualified name of the domain (from the current scope) that is created or a null string if the power domain is not created.	

The **create_power_domain** command defines a power supply distribution network, usually for a list of design elements in the power domain. A *power domain* is a logical grouping of one or more design elements. A power domain shall have one primary power net and one primary ground net. A power domain may have additional supply nets, supply ports, and/or switches.

A power domain is functional only if the primary power and ground supply nets are specified.

–elements (or the **add_domain_elements** command) specifies the complete set of design elements contained within the power domain, i.e., those instances and all their children (unless specified otherwise).

— *list* is specified respective to the current scope; it is not influenced by the **–scope** argument.

— If **–include_scope** is also specified, the scope of the domain is included in the extent of the domain.

— When neither **–elements** nor **–include_scope** is specified, the power domain consists of the current scope and any of its children not specified (as elements) in another **create_power_domain** command.

–scope specifies the scope, i.e., the instance, where the domain shall be created. This scope is the current scope for this command; it defines the domain boundary within the logic design. If **–scope** is not specified, the power domain shall be created within the current scope.

Syntax examples:

```
create_power_domain PD1 -elements {top/U1}

set_scope /top/U1
create_power_domain PD2
```

B.4   create_power_switch

Purpose	Define a switch in the power domain

Syntax	create_power_switch *switch_name*
	    -domain *domain_name*
	    -output_supply_port {*port_name supply_net_name*}
	    {-input_supply_port {*port_name supply_net_name*}}*
	    {-control_port {*port_name net_name*}}*
	    {-on_state {*state_name input_supply_port {boolean_function}*}}*
	    [-on_partial_state {*state_name input_supply_port {boolean_function}*}]*
	    [-ack_port {*port_name net_name [{boolean_function}]*}]*
	    [-ack_delay {*port_name delay*}]*
	    [-off_state {*state_name {boolean_function}*}]*
	    [-error_state {*state_name {boolean_function}*}]*

Arguments		
	switch_name	The name of the switch instance to create; this shall be a simple name.
	-domain *domain_name*	The domain containing the switch.
	-output_supply_port *port_name supply_net_name*	The output supply port of the switch and the net where this port connects.
	-input_supply_port *port_name supply_net_name*	The input supply port of the switch and the net where this port is connected.
	-control_port {*port_name net_name*}	A control port on the switch and the net where this control port connects.
	-on_state {*state_name input_supply_port {boolean_function}*}	A named state, the *input_supply_port* for which this is defined, and its corresponding Boolean function.
	-on_partial_state {*state_name input_supply_port {boolean_function}*}	A named state, the *input_supply_port* for which this is defined, and its corresponding Boolean function.
	ack_port {*port_name net_name [{boolean_function}]*}	A named state, the *input_supply_port* for which this is defined, and its corresponding Boolean function where the switch is in a current-limited state.
	-ack_delay {*port_name delay*}	The acknowledge port on the switch and the signal net where this port connects. Optionally, a boolean function can also be specified.
	-off_state {*state_name {boolean_function}*}	The acknowledge port and delay on the switch where this port connects.

	-error_state {state_name {boolean_function}}	Any error states, which if defined on the switch can be flagged during simulation or analysis.
Return value	Return the full path name of the created switch if successful or a null string if not.	

The **create_power_switch** command defines an instance of a power switch in the power domain. The switch is created within the scope of the power domain. The switch is either on or off.

The switch is on if the value on the control port(s) equals an "on" state Boolean expression; this drives the output port to the "on" state. If a value is specified on the input supply port, then that value is driven on the output supply port.

If the switch is not on, it is off, and an "off" state is driven onto the output port. Some "off" states can be identified as error states. The simulation semantics for these error states is tool dependent.

If a *boolean_function* is specified for **–ack_port**, the result of *boolean_function* is driven on **–ack_port**'s *port_name delay* time units after a control port transition. Otherwise, a logic 1 shall be driven on the *port_name delay* time units after the switch is closed and a logic 0 shall be driven *delay* time units after the switch is opened. *delay* (default is 0) may be specified as a unit-less natural integer or as a Verilog time unit. If specified as a natural integer, the time unit shall be the same as the simulation precision.

Any **–ack_port**, **-on_state**, or **-error_state** *boolean_function*s shall be SystemVerilog Boolean expressions.

The following also apply.

— It is an error if a specified control port does not exist. The existing connectivity of the control ports is unmodified.

— All states not covered by three states (on, off, and error) are anonymous error states.

— Each state name shall be unique.

— Any conflicting state definition is an error.

— No synthesis semantics are associated with any Boolean function, except the mapped switch has to provide that functionality.

Syntax example:

```
create_power_switch sw1
-domain PD_SODIUM
-output_supply_port {vout VN3}
-input_supply_port {vin1 VN1}
-input_supply_port {vin2 VN2}
-control_port {ctrl_small ON1}
-control_port {ctrl_large ON2}
-control_port {ss SUPPLY_SELECT}
-on_state {partial_s1 vin1 {ctrl_small & !ctrl_large & ss}}
-on_state {full_s1 vin1 {ctrl_small & ctrl_large & ss}}
-on_state {partial_s2 vin2 {ctrl_small & !ctrl_large & !ss}}
-on_state {full_s2 vin2 {ctrl_small & ctrl_large & !ss}}
-error_state {no_small {!ctrl_small & ctrl_large}}
```

B.5   create_pst

Purpose	Create a power state table with a specific ordering of supply nets	
Syntax	**create_pst** *table_name* **-supplies** *list*	
Arguments	*table_name*	The power state table name (PST).
	-supplies *list*	The list of supply nets or ports to include in each power state of the design.
Return value	Return the name of the power state table if it is created or the null string if not.	

The **create_pst** command creates a PST, using a specific order of supply nets.

A *power state table* is used for implementation — specifically for synthesis, analysis, and optimization. It defines the legal combinations of states, i.e., those combinations of states that can exist at the same time during operation of the design.

The power state table has no simulation semantics. It is tool-dependent whether simulation tools report an error if an illegal (unspecified) combination of states occurs.

It is an error if a specified supply net has not already been created.

Syntax example:

```
create_pst MyPowerStateTable -supplies {PN1 PN2 SOC/OTC/PN3}
```

B.6   create_supply_net

Purpose	Create a power or ground supply net	

Syntax	**create_supply_net** *net_name*
	    -domain *domain_name*
	    [**-reuse**] [**-resolve** <unresolved | one_hot | parallel>]	

Arguments	*net_name*	The name of the supply net; this shall be a simple (non-hierarchical) name.
	-domain *domain_name*	The domain in whose scope the supply net is to be created.
	-reuse	Extend *net_name* as a supply net within *domain_name*. No new nets are created.
	-resolve <**unresolved** | **one_hot** | **parallel**>	A resolution mechanism which determines the state and voltage of the supply net from the state and voltage values supplied by each of the individual switches. The default is **unresolved**.
Return value	Return the fully qualified name (from the current scope) of the created net or a null string if the net is not created.	

The **create_supply_net** command creates a supply net. The net is defined for the power domain, created in the logic hierarchy at the same scope as *domain_name*, and propagated through implicitly created ports and nets through the logic hierarchy as required .

The following also apply.

— It is an error if *domain_name* does not indicate a previously created power domain.

— When **-reuse** is specified, it is an error if *net_name* does not already exist.

— When **-resolve unresolved** is specified, this supply net only allows a single driver.

Syntax example:

```
create_supply_net v09
    -domain PD1
```

B.7   create_supply_port

Purpose	Create a port on a power domain	

Syntax	**create_supply_port** *port_name*
	    [**-domain** *domain_name*]
	    [**-direction** <**in** | **out**>]	

Arguments	*port_name*	The name of the supply port. Hierarchical names are allowed, unless **-domain** is also specified.
	-domain *domain_name*	The domain where this port defines a supply net connection point.
	-direction <**in** | **out**>	The direction of the port. The default is **in**.
Return value	Return the fully qualified name (from the current scope) of the created port or a null string if the port is not created.	

The **create_supply_port** command defines a supply port at the scope of the power domain when **-domain** is specified, or at the current scope if **-domain** is not specified.

-direction defines how state information is propagated through the supply network as it is connected to the port. If the port is an input port, the state information of the external supply net connected to the port shall be propagated into the domain. Likewise, for an output port, the state information of the internal supply net connected to the port shall be propagated outside of the domain.

It is an error if *port_name* creates a name conflict in the logic hierarchy or specifies a previously created port.

It is an error if *domain_name* does not indicate a previously created power domain.

Syntax example:

```
create_supply_port VN1
    -domain PD1
    reg/wire, Bit, Logic
```

B.8 set_domain_supply_net

Purpose	Set the default power and ground supply nets for a power domain	

Syntax	**set_domain_supply_net** *domain_name*
	    -primary_power_net *supply_net_name*
	    -primary_ground_net *supply_net_name*	

Arguments	*domain_name*	The domain where the default supply nets are to applied.
	-primary_power_net *supply_net_name*	The primary power supply net.
	-primary_ground_net *supply_net_name*	The primary ground net.
Return value	Return 1 if it succeeds and 0 if it fails.	

The **set_domain_supply_net** command associates the default power and ground supply nets to the logic elements of the power domain.

The *primary power and ground nets* are the default nets connected to the logic elements (or inferred cells) of the power domain. These power and ground nets are used for all the elements in the power domain. At the gate level, this means all power/ground pins of all the inferred gates are connected to the primary power/ground nets, unless specified otherwise in a **connect_supply_net**, **set_retention**, or **set_isolation** command.

It is an error if *domain_name* does not indicate a previously created power domain.

It is an error if *domain_name* already contains a primary power and ground net supply.

Syntax example:

```
set_domain_supply_net PD1
-primary_power_net PG1
-primary_ground_net PG0
```

B.9 set_isolation

Purpose	Specify the elements in the domain to isolate using the specified strategy	

Syntax	**set_isolation** *isolation_name*
	    -domain *domain_name*
	    <**-isolation_power_net** *net_name* | **-isolation_ground_net** *net_name* |
	    -isolation_power_net *net_name* **-isolation_ground_net** *net_name* | **-no_isolation**>
	    [**-elements** *list*] [**-clamp_value** <0 | 1 | *latch* | **Z**>] [**-applies_to** <inputs | outputs | both>]	

Arguments	*isolation_name*	Isolation strategy name.
	-domain *domain_name*	The domain for which this strategy is applied.
	<**-isolation_power_net** *net_name* | **-isolation_ground_net** *net_name* | **-isolation_power_net** *net_name* | **-isolation_ground_net** *net_name* | **-no_isolation**>	The supply net(s) used to supply the isolation logic inferred by this strategy. Does not isolate the port, pin, or design element specified in the elements list.
	-elements *list*	A list of design elements, input ports/pins, output ports/pins, and nets to which this strategy is applied.
	-clamp_value <0 | 1 | *latch* | **Z**>	The value to which the input or output shall be clamped. The default is **0**.
	-applies_to <inputs | outputs | both>	Whether the domain's input ports, output ports, or both are isolated. The default is **outputs**.
Return value	Return 1 if it succeeds and 0 if it fails.	

The **set_isolation** command specifies the elements in the domain to isolate using the specified strategy — the isolation enable signal common to all, clamp value, and location.

–isolation_supply_nets can specify a single power net, a single ground net, or both. If only an isolation power net is specified, then the primary ground serves as the isolation ground. If only an isolation ground net is specified, then the primary power net serves as the isolation power.

At least one of **-isolation_power_net** or **-isolation_ground_net** shall be specified, unless **-no_isolation** is specified. If only **-isolation_power_net** is specified, the primary ground net shall be used as the isolation ground supply. If only **-isolation_ground_net** is specified, the primary power net shall be used as the isolation power supply. If both are specified, then these options specify the supply nets to use as the isolation power and ground nets.

The isolation power and ground nets are automatically connected to the implicit isolation processes and, when **-clamp_value** is *latch*, to the isolation value register described in section 5 of the full specification or to the power and ground pins of an isolation cell if an isolation cell is explicitly mapped with the **map_isolation_cell** command. The switched off and on semantics for the implicit isolation process and isolation register, if present, are as described in section 5 of the full specification.

If there are multiple isolation strategies for one domain then, per strategy, **–elements** can be used to specify the elements to isolate. If **–elements** is specified, the elements shall be in the *domain_name*. If **–elements** directly specifies a port by name (not implicitly, by specifying the port's instance or an ancestor of that instance), then the isolation strategy shall apply to that port regardless of whether that port's mode matches the one specified by the **–applies_to** option. When **–elements** is not specified, this is equivalent to using the elements list that defines the power domain.

-clamp_value can be:

— logic 0
— logic 1
— *latch* (the value of the non-isolated port when the isolation signal becomes active)
— logic Z

The following also apply.

— This command never applies to *inout* ports.
— If **-no_isolation** is specified, then all the ports specified in **-elements**, explicitly or implicitly, and matching the **-applies_to** (mode) shall not be isolated.
— It is an error if **-no_isolation** is specified with any other arguments other than - **domain**, **-elements**, or -**applies_to**.
— It is an error if more than one power or ground net is specified.
— It is an error if multiple isolation strategies specify the same design element, pins, ports, or nets.

Syntax example:

```
set_isolation outputs_only
  -domain PD1
  -isolation_power_net VDDbackup
  -clamp_value 1
  -applies_to outputs
```

B.10 set_isolation_control

Purpose	Specify the control signals for a previously defined isolation strategy

Syntax	**set_isolation_control** *isolation_name*
	    -domain *domain_name*
	    -isolation_signal *signal_name*
	    [**-isolation_sense** <high | low>]
	    [**-location** <self | parent | sibling | fanout | automatic>]

Arguments	*isolation_name* Isolation strategy name.
	-domain *domain_name* The domain where the strategy applies.
	-isolation_signal *signal_name* The signal that causes the specified element to drive its clamp value.
	-location <self | parent | sibling | fanout | automatic> Where the isolation cells are placed in the logic hierarchy. The default is **automatic**.
Return value	Return 1 if it succeeds and 0 if it fails.

The **set_isolation_control** command allows the specification of the isolation control signal and sense separate from the **set_isolation** command for those situations where the isolation strategy is known, but the control signals are not known until later.

Excepting the **set_isolation_control** command is executed within the current scope, and the addition of the **–location** option, the semantics here are equivalent to having specified the isolation control signal and sense with the **set_isolation** command.

–location defines where the isolation cells are placed in the logic hierarchy.

   a)   **self** – the isolation cell is placed inside the model/cell being isolated.
   b)   **parent** – the isolation cell is placed in the parent of the cell /model being isolated.

c) **sibling** – a new sibling is created into which the isolation cells are placed.

d) **fanout** – isolation occurs at all fanout locations (sinks) of the port being isolated.

e) **automatic** – the implementation tool is free to choose the appropriate locations (the default).

Syntax example:

```
set_isolation outputs_only
-domain PD1
-isolation_power_net VDDbackup
-clamp_value 1
-applies_to outputs

set_isolation_control outputs_only
-domain PD1
-isolation_signal cpu_iso
-isolation_sense low
-location parent
```

B.11 set_level_shifter

Purpose	Specify a level shifter strategy

Syntax	set_level_shifter *level_shifter_name*
	    -domain *domain_name*
	    [**-elements** *list*] [**-applies_to** <inputs | outputs | both>]
	    [**-threshold** *value*] [**-rule** <low_to_high | high_to_low | both>]
	    [**-location** <self | parent | sibling | fanout | automatic>] [**-no_shift**]

Arguments	*level_shifter_name* Level shifter strategy name (used only for reporting).
	-domain *domain_name* The domain for which this strategy is applied.
	-elements *list* A list of design elements, pins, ports, or nets to which this strategy is applied.
	-applies_to <inputs | outputs | both> Whether the domain's input ports, output ports, or both are level shifted. The default is **both**.
	-threshold *value* The voltage threshold (in volts) for determining when level shifters are required. The default is 0.
	-rule <low_to_high | high_to_low | both> Which type of level shifters are required. The default is **both**.
	-location <self | parent | sibling | fanout | automatic> Where the level shifter is placed in the logic hierarchy. The default is **automatic**.
	-no_shift Can be specified with the **–elements** option to prevent the insertion of level shifters on the specified ports/pins and nets.
Return value	Return 1 if it succeeds and 0 if it fails.

The **set_level_shifter** command can be used to set a strategy for level shifting during implementation. *Level shifters* are the placed on signals that have sources and sinks operating at different voltages, as their associated design elements are connected to different supply nets. If a level shifter strategy is not specified on a particular power domain, the default level shifter strategy consists of all elements in the power domain and uses the default strategy settings.

If **–elements** is specified, the elements shall be in the *domain_name*. If **–elements** is used to specify a port or pin, a level shifter is inserted on that port regardless of any

-threshold or -rule specifications. The -threshold specification defines how large the voltage difference between the driver and sink needs to be before level shifters are inserted. Normally, this threshold is determined from the cell libraries; use this option to override the library values.

–rule can be low_to_high, high_to_low, or both. If low_to_high is specified, signals going from a lower voltage to a higher voltage get a level shifter when the voltage difference exceeds that specified by –threshold. If high_to_low is specified, signals going from a higher voltage to a lower voltage get a level shifter when the voltage difference exceeds that specified by –threshold. If both is specified, it is equivalent to having specified both rules in the strategy.

–location defines where the level shifter cells are placed in the logic hierarchy. All necessary supplies need to be available in the specified location.

a)   Self – the level shifter cell is placed inside the model/cell being shifted.

b)   Parent – the level shifter cell is placed in the parent of the cell /model being shifted.

c)   Sibling – a new sibling is created into which the level shifter cells are placed.

d)   Fanout – level shifter occur at all fanout locations (sinks) of the port being shifted.

e)   Automatic – the implementation tool is free to choose the appropriate locations.

The following also apply.

—   This command never applies to *inout* ports.

—   It is an error if the specified location is not within the logic design starting at the design root.

Syntax example:

```
set_level_shifter shift_up
   -domain PowerDomainZ
   -applies_to outputs
   -threshold 0.02
   -rule both
```

B.12 set_retention

Purpose	Specify which registers in the domain need to be retention registers and set the save and restore signals for the retention functionality

Syntax	**set_retention** *retention_name*
	    -domain *domain_name*
	       <-retention_power_net *net_name* | **-retention_ground_net** *net_name* |
	       -retention_power_net *net_name* **-retention_ground_net** *net_name>*
	    [-elements *list*]

Arguments	*retention_name* — Retention strategy name (used only for reporting).
	-domain *domain_name* — The domain for which this strategy is applied.
	<-retention_power_net *net_name* | **-retention_ground_net** *net_name* | **-retention_power_net** *net_name* **-retention_ground_net** *net_name>* — The supply net(s) used to supply the retention registers inferred by this strategy.
	-elements *list* — A list of objects in the power domain: design elements, named processes, or sequential `reg` or signal names to which this strategy is applied.
Return value	Return 1 if it succeeds and 0 if it fails.

The **set_retention** command specifies which registers in the domain need to be retention registers and identifies the save and restore signals for the retention functionality. Only the registers implied in the elements list shall be provided retention capabilities. If a design element is specified, then all registers within the design element acquire the specified retention strategy. If a process is specified, then all registers inferred by the process acquire the specified retention strategy. If a `reg`, signal, or variable is specified and that object is a sequential element, then the implied register acquires the specified retention strategy. Any specified `reg`, signal, or variable that does not infer a sequential element shall not be changed by this command.

At least one of **-retention_power_net** or **-retention_ground_net** shall be specified. If only **-retention_power_net** is specified, the primary ground net shall be used as the retention ground supply. If only **-retention_ground_net** is specified, the primary power net shall be used as the retention power supply. If both are specified, then these options specify the supply nets to use as the retention power and ground nets.

The retention power and ground nets are automatically connected to the implicit save and restore processes and shadow register or to the power and ground pins of a retention cell when that retention cell is explicitly mapped with the **map_retention_cell** command. The switched off and on semantics for the implicit save and restore processes and shadow register are as described in section 5 of the full specification.

If **-save_signal** is specified, then **-restore_signal** shall be specified. If **-save_signal** is not specified, then **-restore_signal** shall not be specified. If the save and restore signals are not specified, then they shall be specified in a **set_retention_control** command.

If **–elements** is specified, the elements shall be in the *domain_name*. When **–elements** is not specified, this is equivalent to using the elements list that defines the power domain.

The following also apply.

— It is an error if *domain_name* does not indicate a previously created power domain.

— It is an error if more than one power or ground net is specified.

Syntax example:

```
set_retention my_retention
  -domain PDA
  -retention_power_net volt_high
```

B.13 set_retention_control

Purpose	Specify the control signals and assertions for a previously defined retention strategy
Syntax	**set_retention_control** *retention_name* 　**-domain** *domain_name* 　**-save_signal** {{*net_name* <**high** | **low** | **posedge** | **negedge**>}} 　**-restore_signal** {{*net_name* <**high** | **low** | **posedge** | **negedge**>}} 　[**-assert_r_mutex** {{*net_name* <**high** | **low** | **posedge** | **negedge**>}}]* 　[**-assert_s_mutex** {{*net_name* <**high** | **low** | **posedge** | **negedge**>}}]* 　[**-assert_rs_mutex** {{*net_name* <**high** | **low** | **posedge** | **negedge**>}}]*

Arguments	*retention_name*	Retention strategy name (used only for reporting).
	-domain *domain_name*	The domain for which this strategy is applied.
	-save_signal *save_net*	The signal that causes the register values to be saved into the shadow registers.
	-restore_signal *restore_net*	The signal that causes the register values to be restored from the shadow registers.
	-assert_r_mutex {{*net_name* <**high** | **low** | **posedge** | **negedge**>}}	The restore signal for assertion.
	-assert_s_mutex {{*net_name* <**high** | **low** | **posedge** | **negedge**>}}	The save signal for assertion.
	-assert_rs_mutex {{*net_name* <**high** | **low** | **posedge** | **negedge**>}}	Both signals (save and restore) for assertion.

Return value	Return 1 if it succeeds and 0 if it fails.

The **set_retention_control** command allows the specification of the retention control signal and sense separate from the **set_retention** command for those situations where the retention strategy is known, but the control signals are not known until later. As the assertions are related to the save and restore signals, they can also be specified with this command.

Excepting the **set_retention_control** command is executed within the current scope, the semantics here are equivalent to having specified the retention control signals, senses, and assertions with the **set_retention** command.

The **set_retention** command can also be used to specify any assertion options. Each option creates one or more assertions, which verification tools can trigger when the indicated RTL signals are active simultaneously. If **-assert_rs_mutex** does not specify a list of signals, this indicates the save and restore signals themselves are mutually exclusive.

The following also apply.

— The save signal shall be an existing net, port, or pin in the design.
— The restore signal shall be an existing net, port, or pin in the design.

Syntax example:

```
set_retention my_retention_strategy
-domain PDA

set_retention_control my_retention_strategy
-domain PDA
-save_signal {power_controller_inst/save_1 high}
-restore_signal {power_controller_inst/restore_1 low}
-assert_rs_mutex {clock_a posedge}
```

B.14 set_scope

Purpose	Specify the current UPF scope
Syntax	**set_scope** *instance*
Arguments	*instance* — The instance that becomes the current scope upon completion of the command.
Return value	Returns the current scope prior to execution of the command as a full path string relative to the current design top if successful and the null string if it fails (e.g., if the instance does not exist).

If the **set_scope** command is called with no arguments or the UPF scope is not set, the scope is set to the top-level design.

If *instance* is ., the scope remains at the current instance. If *instance* is .., the context is moved up one level in the instance hierarchy. If *instance* begins with /, the scope returns to the instance whose name follows the /, relative to the design top.

Syntax examples:

```
set_scope foo/bar

set_scope ..
```

Glossary

Power Domain: A collection of design elements that share a single primary supply connection and, at least conceptually, share a common power strategy.

Isolation: Isolation is a technique for controlling the behavior of a signal that is driven by a powered down power domain. Isolation consists of driving the signal to a known state - 1, 0, or latching it to a previous value when the power domain is powered down.

Retention: Retention is a technique for retaining the state value of registers in a powered down power domain.

Isolation Cells: Cells (gates) that perform the isolation function in a design. Also knows as clamp cells or clamps.

Level Shifters: Cells (typically buffers) that translate inputs with one voltage swing to an output with a different voltage swing.

Retention Register: A register than extends the functionality of a normal register (flip-flop) with the ability to retain its memory during power down, assuming an appropriate second (always on) supply as well as save and restore signaling.

Shadow Register: The section of a retention register retains the register state during power down. Also known as a balloon registers (due to the topology of some implementations).

Power Switch: At the RTL and architectural level, a power switch allows the power to a power domain to be switched on or off (also known as power gated).

Switching Network: Physically, the power switch is implemented as a switching network of transistors. Also known as a switching fabric.

Switching Transistor: The individual switching transistor that makes up the switching network. Also know as a sleep transistor.

Bibliography

Leakage and Design Methodology

Siva G. Narendra, Anantha Chandrakasan, *Leakage in Nanometer CMOS Technologies*, Springer, 2006

Kaushik Roy, Saibal Mukhopadhyay, and Hamid Mahmoodi-meimand, "Leakage current mechanism and leakage reduction techniques in deep-submicrometer CMOS circuits", Proc. IEEE Vol. 91, No. 2, Feb. 2003

C. Neau and K. Roy "Optimal Body Bias Selection for Leakage Improvement and Process Compensation over Different Technology Generations", Proc. ISLPED, 2003

Dongwoo Lee, David Blaauw, and Dennis Sylvester, "Gate oxide leakage current analysis and reduction for VLSI circuits", - IEEE Trans. VLSI, Vol. 12, No. 2, Feb. 2004

Kiat-Seng Yeo and Kaushik Roy, *Low-voltage, Low-power VLSI Subsystems*, McGraw-Hill, 2005

Enrico Macii, *Ultra low-power electronics and design*, Kluwer Academic Pub. 2004

Jan M. Rabaey and Massound Pedram, *Low power design methodologies*, Kluwer Academic Pub. 2002

Massound Pedram and Jan M. Rabaey, *Power aware design methodologies*, Kluwer Academic Pub. 2002

Kaushik Roy and Sharat C. Prasad, *Low-power CMOS VLSI Circuit Design*, John Wiley &sons, 2000

Gary K. Yeap, *Practical low power digital VLSI design*, Kluwer Academic Pub. 1998

MTCMOS Design, Power Planning and Sleep Transistors

M. Powell, S.-H Yang, et. al., "Gated-Vdd: A circuit technique to reduce leakage in deep-submicron cache memories", in Proc. Int. Symp. Low Power Electronics Design, 2000, pp. 90-95

F. Hamzaoglu and M.R.Stan. "Circuit-level techniques to control gate leakage for sub-100nm CMOS", in Proc. Int. Symp. Low Power Electronics and Design, 2002, pp. 60-63

Satoshi Shigematsu et. al., "A 1-V high-speed MTCMOS circuit scheme for power-down application circuits", IEEE J. Solid-State Circuits, Vol. 32, No. 6, June, 1997

Benton H Calhoun, Frank A Honore and Anantha P Chandrakasan, "A leakage reduction methodology for distributed MTCMOS", IEEE J. Solid-State Circuits, Vol. 39, No. 5, May, 2004, pp. 818-826

Changbo Long and Lei He, "Distributed sleep transistor network for power reduction", Proc. IEEE/ACM Design Automation Conference, 2003

Anand Ramalingam, Bin Zhang, Anirudh Davgan and David Pan, "Sleep Transistor Sizing Using Timing Criticality and Temporal Currents", Proc. ASP-DAC, 2005

Rajat Chaudhry, David Blaauw, Rajendran Panda, Tim Edwards, "Current Signature Compression For IR-Drop Analysis", in Proc. DAC, Los Angeles, California, 2000

J. Kao, A. Chandrakasan and D. Antoniadis, "Transistor sizing issues and tool for multi-threshold CMOS technology", in Proc. Design Automation Conference, 1997

J. Kao, S. Narendra and A. Chandrakasan, "MTCMOS hierarchical sizing based on mutual exclusive discharge patterns", in Proc. Design Automation Conference, 1998

M. Anis, S. Areibi and M. Elmasry, "Design and optimization of multi-threshold CMOS (MTCMOS) circuits", IEEE Trans. Computer-Aided Design of Integrated Circuits and Systems, 2003

M. Powell, S.-H Yang, et. al., "Gated-Vdd: A circuit technique to reduce leakage in deep-submicron cache memories", in Proc. Int. Symp. Low Power Electronics Design, 2000, pp. 90-95

Anand Ramalingam, Bin Zhang, Anirudh Davgan and David Pan, "Sleep Transistor Sizing Using Timing Criticality and Temporal Currents", Proc. ASP-DAC, 2005

Changbo Long and Lei He, "Distributed sleep transistor network for power reduction", Proc. IEEE/ACM Design Automation Conference (DAC), 2003

Benton H Calhoun, Frank A Honore and Anantha P Chandrakasan, "A leakage reduction methodology for distributed MTCMOS", IEEE J. Solid-State Circuits, Vol. 39, No. 5, May, 2004, pp. 818-826

Changbo Long, Jinjun Xiong and Lei He, "On optimal physical synthesis of sleeper transistors", Proc. ISPD, 2004

Kaijian Shi, Zhian Lin, Yi-min Jiang, "A Power Network Synthesis Method for Industrial Power Gating Designs", Proc. IEEE Int. Symposium on Quality Electronic Design (ISQED) March, 2007

Kaijian Shi and Jason Binney, "Design Optimization Methodologies for Low-Leakage Power Designs in Sub-90nm Technology", Proc. Euro DesignCon, 2004

Kaijian Shi and David Howard, "Challenges in Sleep Transistor Design and Implementation in Low-Power Designs", Proc. IEEE/ACM Design Automation Conference (DAC) 2006

Kaijian Shi and David Howard, "Sleep Transistor Design and Implementation – Simple Concepts Yet Challenges To Be Optimum", Proc.. IEEE VLSI-DAT, April, 2006

John Biggs and Alan Gibbons, "Aggressive leakage management in ARM based systems", Proc. Synopsys Network of User Group (SNUG), Boston, 2006

Memory Design for Low Power

Chris Hyung-il Kim, Jae-Joon Kim, Saibal Mukhopadhyay, Kaushik Roy, "A forward body-biased low-leakage SRAM cache: device, circuit and architecture considerations", IEEE Trans. VLSI, Vol. 13, No. 3, 2005

Yasuhisa Takeyama, Hiroyuki Otake, Osamu Hirabayashi, Keiichi Kushida, Nobuaki Otsuka, "A low leakage SRAM macro with replica cell biasing scheme", IEEE J. Solid-State Circuits, Vol. 41, No. 4, 2006

Kyeong-Sik MIN, Kouichi KANDA, Hiroshi KAWAGUCHI, Kenichi INAGAKI, Fayez Robert SALIBA, Hoon-Dae CHOI, Hyun-Young CHOI, Daejeong KIM, Dong Myong KI2 and Takayasu SAKURAI, "Row-by-Row Dynamic Source-Line Voltage Control (RRDSV) Scheme for Two Orders of Magnitude Leakage Current Reduction of Sub-1-V-VDD SRAM's", IEICE Trans. on Electronics, Vol E88-C, No. 4, pp 760-767, 2005

A. Agarwal, H. Li, and K. Roy, "A Single-Vt Low-Leakage Gated-Ground Cache for Deep Submicron," IEEE J. Solid-State Circuits, IEEE Press, 2003, pp. 319-328

A. Agarwal and K. Roy, "Noise Tolerant Cache Design to Reduce Gate and Sub-threshold Leakage in Nanometer Regime," Proc. Int'l Symp. Low Power Electronics and Design (ISLPED 03), 2003, pp. 18-21

C.H. Kim and K. Roy, "Dynamic Vt SRAM: A Leakage Tolerant Cache Memory for Low Voltage Microprocessors," Proc. Int'l Symp. Low Power Electronics. and Design (ISLPED 02), ACM Press, 2002, pp. 251-254

C.H. Kim et. al., "A Forward Body-Biased Low-Leakage SRAM Cache: Device and Architecture Considerations," Proc. Int'l Symp. Low Power Electronics and Design (ISLPED 03), ACM Press, 2003, pp. 6-9

K. Flautner, "Drowsy Caches: Simple Techniques for Reducing Leakage Power," Proc. 29th Ann. Int'l Symp. Comp. Architecture (ISCA-29), IEEE CS Press, 2002, pp. 148-157

S. Heo et. al., "Dynamic Fine-Grain Leakage Reduction Using Leakage-Biased Bit-lines," Proc. Int'l Symp. Comp. Architecture(ISCA- 29), IEEE CS Press, 2002, pp. 137-147

K. Itoh et. al., "A Deep Sub-V, Single Power-Supply SRAM Cell with Multi-Vt, Boosted Storage Node and Dynamic Load," Proc. Symp. VLSI Circuits Digest of Technical Papers, IEEE Press, 1996, pp. 132-133

Index